Careers in Nursing and Related Professions

LINDA NAZARKO

KOGAN PAGE

To my husband Ed, for his support,
as always.

First published in 1979 entitled *Careers in Nursing and Allied Professions*
Second edition 1983, reprinted 1984
© Stephen Moss 1979, 1983
Third edition 1985, reprinted 1986
© Rosemary Clark and Stephen Moss 1985
Fourth edition 1989
Fifth edition 1991
© Rosemary Clark
Sixth edition 1994 entitled *Careers in Nursing and Related Professions*;
reprinted with revisions 1995
Seventh edition 1997
© Linda Nazarko

Apart from any fair dealing for the purposes of research or private study or criticism or review, as permitted under the Copyright, Designs and Patents Act, 1988, this publication may only be reproduced, stored or transmitted, in any form or by any means, with the prior permission in writing of the publishers, or in the case of reprographic reproduction in accordance with the terms of licences issued by the Copyright Licensing Agency. Enquiries concerning reproduction outside those terms should be sent to the publishers at the undermentioned address:

Kogan Page Limited
120 Pentonville Road,
London N1 9JN.

© Linda Nazarko 1994, 1997

British Library Cataloguing in Publication Data

A CIP record for this book is available from the British Library.
ISBN 0-7494-2107-X

Typeset by DP Photosetting, Aylesbury, Bucks
Printed and bound in Great Britain by
Clays Ltd, St Ives plc

Contents

Introduction 1

1. What is Nursing? 3
Who Can Apply for Nurse Training? 3; Is Nursing the Career for You? 3; Types of Training – Diploma or Degree 4; Entry Qualifications 5; The DC Test 6; Branch Programmes 6; Adult Nursing 7; Children's Nursing 7; Learning Disability Nursing 8; Mental Health Nursing 9; Training Allowances 11; Income Support and Housing Benefit 12; Council Tax 12; Becoming a Student 12; Accommodation 13; Applying for Nurse Training 13

2. Midwifery 17
Working as a Midwife 17; Training 18; Pay and Conditions 19; Working as a Qualified Midwife 20; Working as a Community Midwife 20; Refresher Courses 20; Returning to Practice 21; Further Education for Midwives 21

3. Working as a Qualified Nurse 23
Primary Nursing 23; Induction Courses 24; Teaching and Supporting Colleagues 25; Career Development 25; Career Options 26; Working Conditions 27; Compulsory Further Education for Qualified Nurses 30

4. Post-Registration Qualifications 31
Sick Children's Nursing 32; Mental Health Nursing 33; Learning Disability Nursing 34; Adult Nursing 34; Enrolled Nursing 35

5. Continuing Education for Registered Nurses 40
UKCC Areas of Relevance 40; Your PREP Requirements 41; Diploma and Degree Studies 42; CATs Points and Continuing Education 43; Making Your Learning Count 43; Diploma-level Studies 43; Post-Registration Courses 44; Degree-level Studies 45; Master's Degree Studies 45; Combining Theory With Practice: the ENB Higher Award 45; Working as a Nurse Teacher 46; Further Reading 46

iv Contents

6. **Returning to Nursing** 47
   Refresher Courses 47; Supervised Practice 47; Flexible
   Working Hours 47; Arranging Childcare When You Return
   to Work 48; Taking the Plunge 48; Single Parents 49;
   Further Reading 50

7. **Working in the Community** 51
   Community Nursing 51; Specialist Community Nurses 52;
   Community Psychiatric Nursing 53; Health Visitors 53;
   School Nursing 53; Practice Nursing 54

8. **Working in the Independent Sector** 56
   Working in Private Hospitals 56; Elderly Care in the
   Independent Sector 56; Continuing Your Professional
   Development 57; Careers in Nursing Homes 58

9. **Nursing Abroad** 59
   Cultural Differences 59; Career Planning 60; Working in
   the United States 60; Working in Canada 61; Working in
   Australia 62; Working in Saudi Arabia 63; Working in
   Europe 64; Voluntary Work Abroad 65

10. **Nursery Nursing** 67
    Training 67; Working as a Nursery Nurse 68; Further
    Education 71; Career Options 71

11. **Careers Related to Nursing** 73
    Operating Department Assistant 73; Ambulance Work 73;
    Chiropody 74; Dentistry 75; Dietetics 78; Nurse's Aide 79;
    Occupational Therapy 80; Physiotherapy 81; Speech and
    Language Therapy 82; Radiography 83; Social Work 84;
    Therapy Aides 85

12. **Training Courses** 86
    Pre-Registration Courses Available in England at
    Approved Institutions 86; Diploma-level Pre-Registration
    Nursing Programmes in Northern Ireland 89; Diploma-
    level Programmes in Wales 89; Diploma-level Programmes
    in Scotland 90; Midwifery Programmes in England at
    Approved Institutions 91

Introduction

When I wrote *Careers in Nursing* in 1994 there were two levels of registered nurse and the first diploma-trained nursing students had not qualified. Nurses and midwives were trained at hospital-based schools. Now enrolled nurse training has been discontinued and diploma-level training is the minimum standard for newly registered nurses and midwives. A tenth of all nursing students are combining registered nurse training with a degree. The training of nurses, midwives and health visitors has been replaced by a broader, university-based education. There are predictions that by the turn of the century all newly qualified nurses, midwives and health visitors will be educated to degree level.

Registered nurses, midwives and health visitors must now continue their professional development if they wish to continue practising. Those who have a break in practice will, in future, have to complete a return to practice course before resuming work.

The 1990s have seen enormous changes in health care. A quarter of nurses and growing numbers of people in related careers now work in the independent sector. Increasingly, the focus of nursing is not in hospitals but in the community. The seventh edition of *Careers in Nursing and Related Professions* aims to help people considering a career in the caring professions and offers practical advice to individuals considering further education or a return to practice.

Nursing is still often thought of as a career for young people who have just left school or college. Because of this, many mature people feel that they have left it too late to fulfil their ambition to train as a nurse. Fortunately, nothing could be further from the truth – nurse training is open to people from all walks of life. School-leavers are welcome but people can choose nursing as a second career or train after bringing up a family. Mature students without academic qualifications are welcomed, as are students with degrees in unrelated subjects. In recent years nursing has succeeded in attracting a wider range of people than ever before. A third of all nursing students are over 25 years old when they begin their nursing education; a quarter have dependent children and the number of male students is increasing.

Training Places

The number of student places has fallen in recent years and there is a shortage of registered nurses, but no reports at present of shortages of midwives or health visitors. There are now 12 applicants for every training place. In 1995 18,500 students commenced nurse training. An estimated 80–90 per cent will successfully complete their courses.

Students on diploma-level courses normally have to decide which part of nursing they will specialise in after they commence training. Information is given on each of the specialist branches to help you choose.

The number of pre-registration degree students is growing and this book gives information about degree and diploma programmes so that you can decide which to apply for.

Nursing offers the newly qualified nurse an enormous, sometimes bewildering, range of options. This book will help them to choose which area to specialise in. It discusses careers in hospital and community settings, staff nurse development courses and further education options.

Nursing is a flexible career; nurses bringing up a family may decide to take a break in practice while their children are young. Others may wish to continue working on a part-time or job-sharing basis. Some may decide to move to another area of practice – popular options include working as practice nurses or moving to the independent sector. *Careers in Nursing* provides information about these ideas and about returning to practice, including details of refresher courses.

The discontinuance of enrolled nurse training has led many enrolled nurses to take stock of their careers and to contemplate undertaking a conversion course leading to registered nurse qualifications. This book gives details and explores the advantages and disadvantages of full- and part-time conversion courses and 'open learning' conversion courses.

Changes in nurse education include the introduction of Post-Registration Education and Practice (PREP) legislation which requires all qualified nurses to undertake the equivalent of five study days every three years if they are to continue to practise. Study options including open learning are discussed. Diploma-level nurse education will, within the next few years, mean that many qualified nurses will be registered nurses with a diploma qualification. Many registered nurses trained under the traditional system are keen to obtain diploma qualifications or degrees. The recently introduced national boards' (for England, Scotland, Wales and Northern Ireland) Higher Award schemes linking professional and academic study are discussed. This book explains how nurses can gain credit for previous learning and continue their studies to diploma or degree level.

Many care assistants are now keen to obtain National Vocational Qualifications (NVQs). *Careers in Nursing and Related Professions* provides information for care assistants who wish to obtain NVQs. Other related careers such as ambulance and paramedic work, occupational therapy, physiotherapy, art therapy, social work, work as an operating department assistant are described.

The majority of nurses are women and for this reason I have chosen to refer to nurses as 'she' but this is for convenience only; most nursing posts are open to men and women.

Linda Nazarko, August 1996

Chapter 1
What is Nursing?

Nursing involves the 'hands, heart and head'. It is a demanding career and nurses work shifts providing care for those who need it 365 days a year. As a nurse, you will often be required to work shifts and may have to start work at 7 or 7.30 am on an early shift or finish work at 9.30 or 10 pm on a late shift and to work on night duty. Nurses often have to work at weekends and on bank holidays. Nursing requires intelligence, compassion and stamina.

The rewards are enormous: the chance to assist a qualified midwife as she helps a woman to deliver her child; the opportunity to help an individual learn to walk again following an accident and an operation; the satisfaction of knowing that you have helped a mentally handicapped person to acquire a new skill which leads to greater independence. These are just a few of the many rewards nurses have every day in a career where no day is the same as the one before and where a trained nurse can make a difference to people's lives.

Who Can Apply for Nurse Training?

Nurse training is open to people in all walks of life. Nursing students with experience in other careers or experience of bringing up a family can contribute much to nursing. The woman with children of her own can understand the distress parents feel when their children are ill in hospital and draw on her own experience when comforting them. The adult who has lost her own parents can understand the grief that adults suffer when their parents are ill or dying and can draw on that experience to provide professional, highly skilled care to patients and their families.

Age Limits
Nurses can begin training from the age of $17\frac{1}{2}$ although prospective students can apply from the age of $16\frac{1}{2}$. There is no upper age limit and the maturity and life experience of older students is valued.

Is Nursing the Career for You?
Estimates of the number of students who fail to complete their studies vary from a tenth to a third. It is important to find out as much as you can about nursing so that you can decide if it is right for you. Many colleges of

nursing now look for evidence that the applicant has some knowledge of what nursing involves. Applicants who are aware of the physical and intellectual demands of nursing stand a greater chance of obtaining a place.

Find out about the nature of nursing

Working in a voluntary or paid capacity in a hospital or community setting will help you decide if nursing is the career for you. Many schools now have community service programmes for fifth and sixth formers. If you are still at school ask for a placement in a caring environment to be arranged. Suitable placements include hospitals, nursing homes, day hospitals and day centres.

Individuals who have left school have a number of options:

1. *Working as a volunteer.* Volunteers are welcome in many care settings including hospitals and nursing homes. If you are working you can undertake voluntary work at weekends or in the evening. If you have school-age children you can work for a few hours while they are at school.
2. *Applying for a post as a care assistant.* You may be able to obtain work as a care assistant at the local hospital or nursing home. Many employers prefer to employ experienced care assistants, though some are prepared to offer training.
3. *Applying for a National Vocational Qualification course.* NVQ training includes clinical placements, academic study and assessment of competence. If you have been unemployed for six months or more you may be eligible for free NVQ training at your local college of further education.

The Advantages of Gaining Practical Experience in Caring

Working as a care assistant alongside registered nurses will help you to gain a realistic idea of the nature of nursing work and which type of nursing you would like to enter. It also has one other advantage: applicants for nurse training must supply references. Nurse teachers selecting students will find a reference from a registered nurse more valuable than that of a teacher or non-nursing employer.

Types of Training – Diploma or Degree

We live in a rapidly changing society, where people are living longer; a society where the birth rate is falling and there are more old people than ever before. Developments in science and technology have increased demands for health care and the length of time people spend in hospital has decreased. There is a greater emphasis on community care and a new awareness that nurses have an important part to play, not only in nursing the sick but in health education.

In an effort to meet the demands of our rapidly changing society a new form of nurse education was introduced a few years ago. Nurses now study at diploma level and gain a Diploma in Higher Education as well as a nursing qualification.

Diploma-level Pre-registration Courses (Project 2000)
Diploma-level courses aim to train nurses who are able to combine theory and practice. Diploma students spend more time on theoretical work than traditionally trained nurses. The first 18 months of the three-year course is known as the *Common Foundation Programme*. All diploma students, regardless of their area of specialism, share this 18-month course.

After the common foundation part of training is completed students enter the preparation for their specialist branch. Diploma courses stress the importance of health education and preventing ill health. Because nursing has moved from a hospital-based service to a balance of hospital and community care, students spend more time in the community than previously. Diploma students are allocated to clinical areas to gain experience. They are, in theory, not counted as part of the nursing workforce and have 'supernumerary status'. However, the nursing press is full of reports from students who find that their supernumerary status is threatened by shortages of registered nurses. Students undertake 1,000 hours of rostered service during the third year of training and are counted as part of the unit's staff for this period.

It is too early to decide if diploma training will equip nurses with the skills needed to nurse in the twenty-first century. Project 2000 courses have much larger classes than traditional courses and some nurse teachers are concerned about this. The amalgamation of colleges of nursing into the university sector is now almost complete. It has been claimed that diploma-level courses offer more flexible working hours and that this is helpful to students with young children. This increased flexibility is possible because students are no longer counted as part of the nursing workforce for most of their courses. Some nurse teachers feel that diploma students have too little clinical experience on qualification and many colleges of nursing are now working to provide intensive clinical support.

Degree Courses
Degree programmes are available in all four branches of nursing described on pages 6-7. They vary in length between three and four years full time. Degree students study at university and have clinical placements to gain experience, as diploma students do.

Degree programmes tend to cater for smaller numbers of students and competition for places is more intense than for diploma places. Ten per cent of all nursing students are now undertaking degree programmes.

Entry Qualifications
Now that nurse education is university based, the normal *minimum* entry requirements include:

- Five GCSE passes at grade C or above; some subjects such as needlework or home economics may not be accepted. Normally, GCSEs should have been obtained at one or perhaps two sittings.

- Some vocational qualifications such as BTEC or RSA II and III may count towards entry.
- National Vocation Qualifications at level 3 in either caring or other subjects are increasingly accepted as entry criteria. It is worth ringing up your local college of nursing to check.
- Access courses; details of these can be obtained from your local college of further education.

Nurse training courses are more academically demanding than ever before and if you are still at school you would be wise to sit A levels. The subjects studied are not important unless you are applying for a degree course; it *is* important that they demonstrate that you have the ability to undertake academic studies at a higher level. Some post-registration courses such as health visiting are very popular and students with A levels have a greater chance of obtaining a place on them. If you intend to apply for a combined RN/degree course be warned there are many more applications than there are places available. A level subjects such as biology, chemistry, physics or English greatly improve a candidate's chances of acceptance on such a course.

The DC Test

If you are a mature person keen to train as a nurse do not despair. If you have no educational qualifications or if your qualifications do not meet the entry requirements, it is possible to take an entrance test known as the DC test. This is a test of English, maths and verbal reasoning. The maths section requires a knowledge of fractions and decimals. The test lasts for one and a quarter hours and is timed. The pass mark is 51 per cent minimum – this varies from college to college. Many colleges require a higher pass mark. A pass indicates that you will be able to cope with the academic requirements of the course.

The DC test is offered at the discretion of the training school approached. Many colleges now demand evidence of recent study before they will consider offering a DC test. Evidence of work experience in a caring role, such as working as a nursing auxiliary, in a voluntary capacity in a community or hospital setting or with an organisation such as the Red Cross or St John Ambulance, will be an asset. This type of experience indicates that you have some idea of the role of a nurse. Unsuccessful candidates are allowed a further two attempts but will not take the same paper twice. Individuals considering the DC test will find *Taking the DC Test – A Guide for Candidates* useful. It costs £1.25 and can be obtained from: The Nurse Selection Project (Books), University of Leeds, Leeds LS2 9TJ.

Branch Programmes

Registered nurses are specially trained to care for different groups of patients, and four pre-registration programmes are offered. These are:
- Adult nursing

- Children's nursing
- Learning disability nursing
- Mental health nursing.

Applicants must decide which branch of nursing they wish to qualify in before applying for training. This is most important. Universities have contracts with hospitals to train a certain number of nurses with each type of qualification. It is sometimes possible to move from one branch to another while undertaking the common foundation part of training but only if another student also wishes to swap. Details of the four branches are given below.

Adult Nursing

Nurses are trained in all aspects of adult nursing and spend time not only working on hospital wards but also in outpatients departments, operating theatres, intensive care, wards specially for nursing sick children and elderly people. Students also spend time working with trained midwives caring for mothers before and after childbirth and assisting with the care of newborn babies. They work in the community under the supervision of district nurses, health visitors and community midwives as well as spending time in psychiatric units working under the supervision of registered mental nurses. However, the main focus of their training is adult nursing.

Student nurses are taught the importance of team work throughout their training. The student nurse learns how to use her initiative and to remain alert and aware of possible complications which patients may develop. She learns how to identify any problems which the patient may face as a result of illness, to plan care and to check if she has succeeded in meeting the patient's needs.

After registration nurses often go on to train in another area such as children's nursing. Many senior positions require a general and a specialist qualification. A nurse manager in charge of a gynaecology and maternity unit might require qualifications in adult nursing and midwifery.

Training Opportunities

Pre-registration programmes are available at diploma and degree level. The diploma level lasts three years. Degree programmes take three or four years to complete. Graduates with a relevant health-related degree can apply for accelerated programmes – normally two years long. A shortened course is available for registered nurses with other qualifications; for further information see Chapter 4.

Children's Nursing

Children's nursing is not the same as caring for adults. The child's world centres around home and family. Sick children are normally cared for on special children's wards where the atmosphere is very different from an adult ward. Children can find a stay in hospital frightening and bewildering and younger children especially are unable to express their fears. Children's nurses need to be kind and patient and to give their young patients lots of attention. Advances in healthcare have resulted in children staying in hospital for shorter periods. Now children's nurses are as likely to be working in the community as in hospital.

Children's nurses are specially trained in caring for sick children, and training covers community care, surgical nursing, medical nursing, caring for children with physical and learning disabilities. The children's nurse needs to be sensitive enough to understand the often unspoken fears of children.

Parents can be very upset by their child's illness as can the child's brothers and sisters. The sick children's nurse needs to be able to provide support not only for her young patient but also for the family. Parents are encouraged to stay with their children when they are in hospital and often assist the nurse. When parents are unable to stay, perhaps because they live a great distance away or have no one to care for the other children in the family, the nurse must try to be a substitute parent as well as a nurse.

Communication skills are a very important asset to the children's nurse; she must communicate with other members of the hospital team in a professional way, comfort worried parents and explain technical language so that they will understand treatment; she must also be able to enter into the child's world, comfort, care for, play with and cuddle the children in her care.

The very special thing about children's nursing is that although children can be very ill on admission to hospital they recover much more quickly than adults. Nursing a very sick child one day and returning the next to a laughing, happy, recovering child is a wonderful experience, as any children's nurse will tell you.

The Role of the Children's Nurse
This includes:

- Caring for newborn babies and teenagers.
- Supporting the child's family and helping them to care for their child.
- Working with the child and the family to plan care.
- Working as part of a professional team.
- Assuming responsibility for a child's care.

Training Opportunities
As for adult nursing; see page 7.

Learning Disability Nursing

In recent years the term 'mentally handicapped' has been replaced by the more accurate and sensitive term 'people with learning disabilities'. This reflects the change in attitude that has taken place. It is now realised that people with learning disabilities have a wide range of abilities. In the past, their potential was not always recognised and the emphasis was on caring for them as patients. Now the role of nurses caring for people with learning disabilities has changed and qualified nurses work with patients to enable them to reach their fullest potential.

The nurse assesses each individual and works with the person to develop a plan of care. Some people need to be taught basic skills such as bathing and the nurse breaks each skill into a series of small tasks and works with individuals, helping them to learn to care for themselves. People with learning disabilities often have physical handicaps and nurses work with other professionals such as physiotherapists, occupational therapists and doctors in providing care and promoting independence. The skills of the nurse (the person who has most contact with the patient) are of vital importance. Some individuals with mild degrees of learning disability are helped to learn how to shop and cook. They may live in small groups in hostels in the community where they are able to lead a more normal life than was possible in the large mental handicap hospitals. More people with learning disabilities are cared for at home and in the community than ever before and increasing numbers of nurses are undergoing specialist training so that they can offer care in the community, dealing with people's problems when they arise and reducing the need for hospital admissions.

The Role of the Learning Disability Nurse

Learning disability nurses are now mainly based in the community. The role includes:

- Identifying factors that affect the health and well-being of the individual.
- Working with the person to help him or her to develop the ability to maintain and improve health.
- Working as part of a team, helping the individual to develop new skills and abilities.
- Teaching others about learning disability.

Training Opportunities

Pre-registration programmes are available at diploma and degree level. The diploma level lasts three years. Degree programmes take three or four years to complete. Graduates with a relevant health-related degree can apply for accelerated programmes – normally two years long. Joint training: a few universities offer a three-year course leading to a joint award in nursing and social work. A shortened course is available for registered nurses with other qualifications; for further information see Chapter 4.

Mental Health Nursing

Mental health nurses care for people who are mentally ill. There are many different types of mental illness and many physical and psychological causes for illness. Mental illness is very common and increasing numbers of mentally ill people are being cared for at home or in small homes in a community setting. Some people require treatment in hospital but this is usually short term; over half the patients admitted to a psychiatric unit return home after a month. Treatment is then continued in the community and is supervised by specialist mental health nurses called Community Psychiatric Nurses (see Chapter 7 for further details). The majority of patients with a psychiatric illness can be successfully treated and return to a normal life.

As people are living longer there has been an increase in the numbers of elderly people who become mentally ill. Mental health nurses care for them in a variety of settings: their own homes, special sheltered accommodation, residential homes, nursing homes and specialist units for elderly mentally infirm people, known as EMI units. Elderly mentally infirm people often have physical as well as mental health problems and for this reason many psychiatric nurses also qualify as general nurses.

Mental health nurses care for patients suffering from a wide range of illness. Some patients simply require help, counselling and support from the nurse and are able to solve their own problems. Some develop fears or phobias such as agoraphobia (the fear of open spaces) and require help to overcome these fears and lead a normal life. Many psychiatric patients are aware that they need help and they are diagnosed as suffering from a neurosis. An example of a neurosis would be a fear of flying which might prevent a person from visiting her family in another country. Other patients are more ill and are not aware that they behave abnormally; they are usually diagnosed as suffering from a psychosis. Schizophrenia is an example. Patients suffering from this disease can have difficulty in distinguishing between fantasy and reality and may have hallucinations. In the last 40 years there have been major advances in the drugs available to treat psychotic patients and many make a full recovery in a reasonably short time. Community psychiatric nurses visit patients who have suffered from a psychotic illness when they return home; they observe the patient and can often provide support which reduces the likelihood of relapse.

Mental health nurses require skill, tact and patience in caring for people, but find that helping a patient to recover and resume a normal life is very rewarding.

The Role of the Mental Health Nurse
This can include:

- Assessing the individual's needs, working with the patient to develop a plan of care. Monitoring the effectiveness of planned care.
- Administering and monitoring the dosage of prescribed medicine. Explaining the effects of medication to the individual.
- Working as part of a team of professionals.
- Acting as a 'key worker' co-ordinating the care of an individual.

Training Opportunities
As for adult nursing; see page 7.

Training Allowances

Diploma students, unlike other students, receive a bursary not a grant. Bursaries are not subject to national insurance contributions, income tax or superannuation contributions. The present rates (April 1996) are:

Table 1. Bursaries for Diploma Students

Students aged under 26, Annual Bursary	Students aged over 26, Annual Bursary
£4,450 (outside London)	£5,010 (outside London)
£5,230 (in London)	£5,785 (in London)

Additional allowances are paid to students with dependants. The annual rates (April 1996) are:

Table 2. Dependant Allowances for Diploma Students

Spouse, adult dependant or first child	£1,665
Children under 11	£350
Children aged 11-15	£700
Children aged 16-17	£925
Children aged 18 and over	£1,330
Additional allowance for single students with dependants	£825

Diploma students have seven weeks' holiday a year and do not normally work on bank holidays. They may work weekends, evenings and night duty as part of their 1,000 hours' practical experience but, unlike registered nurses, are not paid additional allowances for working unsocial hours. Diploma students are, in principle, exempt from council tax.

Degree students should apply to their local education authority (LEA) for a grant. Grant regulations are complex and change each year. Table 3 gives details of a typical London university annual maintenance grant (September 1996).

Table 3. Annual Maintenance Grant (London)

Living away from home	£2,795
Living in parental home	£1,769

Additional allowances are paid to students with dependants. The annual rates are (September 1996): Please note these are for guidance only as rates vary.

Careers in Nursing and Related Professions

Table 4. Dependant Allowances for Degree Students

Spouse, adult dependant or first child	£1,865
Children under 11	£395
Children aged 11-15	£785
Children aged 16-17	£1,035
Children aged 18 and over	£1,495

Income Support and Housing Benefit

Full-time students are not normally entitled to income support and housing benefit. Exceptions are:

- Single parents with a dependent child under 16
- Disabled students
- One of a couple with dependent children under 16 where the partner is not in full time study
- Couples who are both students and have dependent children under 16 can claim income support and housing benefit during the summer vacation only.

Council Tax

The rules relating to council tax are complex. In principle, full-time students are exempt from council tax. In practice, this depends where you live and who you live with.

- Students who live in halls of residence are exempt.
- Students sharing with other students are not liable.
- Students who share with non-students are usually liable to pay council tax.

Your chosen university can give you further advice.

Becoming a Student

The decision to apply for a place and begin nurse training is a major one. Many school leavers may still be living at home and the decision may involve moving away from home for the first time, leaving family and friends and going to a new area. You may decide to move to a London teaching hospital or another large city. Find out as much as you can about the colleges you are considering. Consider leisure facilities on offer. Life as a student nurse is not all work and no play. It also makes sense to check on facilities for study such as well-stocked libraries and access to computers and computer searching facilities.

If you are a mature student with domestic responsibilities you may be unable or unwilling to move across the country to train. You may have a different set of priorities. You may wish to enquire about the availability of nursery places so that your child can be cared for while you study. Nursery places are highly sought after and waiting lists are common; single parents and students who would be unable to study without childcare are usually given priority. Fees are on a sliding scale and related to income. You may want to find out about holiday play schemes for your school-age children. You may wish to enquire about the possibility of flexible working hours during your clinical placements.

Most colleges of nursing produce their own brochures which give details of facilities available. You can obtain brochures by writing to the hospitals you are interested in. Many have open days when you have an opportunity to look around the hospital and college of nursing. You will probably be able to chat to students and newly qualified nurses over coffee and have an opportunity to ask questions.

Accommodation

In the past, student nurses, apart from mature students, were required to live in hospital accommodation and could apply for permission to 'live out' after the first year of training. Now hospital accommodation is in short supply and student nurses are usually offered hospital accommodation for the first six months of training only. This enables them to settle into a new area and find their own flat or house. Hospitals, colleges of nursing and universities often keep lists of flats and houses which are usually shared by a number of students to keep costs down. Staff notice boards in colleges of nursing and hospitals may offer a large selection of accommodation. Project 2000 students are normally expected to find their own accommodation.

You may feel very alone and homesick in the first few months of training but this is normal and you will adjust to this new situation and make new friends with whom you will probably keep in touch long after your basic training is completed.

Mature students will face a major period of readjustment; research shows that mature students with families settle into nursing and cope well academically, often outshining their younger colleagues.

Applying for Nurse Training

England

Diploma Programmes
Nurses applying for training in England now apply to the Nurses and Midwives Central Clearing House; this is a department of the English National Board (ENB) and provides a centralised system which deals with all applications for *full-time* courses. Details of courses available are given in Chapter 12.

Applications for Courses
Applications for places on programmes commencing in autumn 1997 had to be in before 13 December 1996.

Individuals wishing to apply in 1997 will be able to obtain the application package in May. The clearing house procedure will transfer to a department managed by the Universities and Colleges Admissions Service (UCAS) in 1997 called the Nursing and Midwifery Admissions Service (NMAS). The application package will be available from:

NMAS, Fulton House, Jessop Avenue, Cheltenham, Gloucestershire GL50 3SH.

The Application Procedure

An information leaflet is sent out and it should be returned with a cheque for £10. You then receive an application handbook which contains a guide to all the schools of nursing in the country you wish to train in and lists the normal entry requirements and the courses offered. This guide is also available in local reference libraries. You also receive an application form. You can make up to six choices of training college and copies of your application are sent to the first three on your list. The colleges of nursing then usually contact you directly and can invite you to open days or arrange interviews.

You will probably have to complete a health questionnaire and have a medical on the same day as the interview. Colleges of nursing have different policies after the interview: some inform applicants of the result of the interview on the day; others write to applicants within a few days of the interview giving the result. Full details of the application procedure are in the handbook.

Degree Programmes

Individuals wishing to apply for degree programmes should contact:

UCAS, Fulton House, Jessop Avenue, Cheltenham, Gloucestershire GL50 3SH.

Part-time Courses

Applicants for part-time courses should enquire about their availability locally and will be given advice on local application procedures.

Further Information

Further information on courses in England can be obtained from:

ENB Careers, PO Box 2EN, London W1A 2EN.

You can telephone the ENB (Monday to Friday between 10 am and 4 pm) on 0171 391 6200 or 0171 391 6205.

If you have access to a computer with a modem you can obtain information about nursing careers from ENB via the Internet: Internet http://www-enb.org.uk

You may find the nursing journals helpful for gaining an understanding of the issues affecting nursing in the 1990s. This knowledge could be invaluable during your interview. The most popular nursing journals are: *Nursing Times* and *Nursing Standard and Professional Nurse* which can be bought from any large newsagent.

Northern Ireland

Diploma Courses
Northern Ireland currently has four main colleges of nursing: Northern, Southern, Eastern and Western Colleges of Nursing, and there are plans to integrate the colleges with the university sector. Currently, applicants should apply to the individual colleges of nursing. Details are given in Chapter 12.

Degree Courses
The University of Ulster offers a degree course which lasts four years. Students have the opportunity to learn a European language. They undertake a 12-week overseas research placement. Only 30 places are available each year. Entry criteria are stringent: normally three A levels – two B grades and one C grade – are required. The University prefers to accept students with one science subject. O level or GCSE passes in English language and mathematics are also required. Application is through UCAS.

Further information can be obtained from:

Mr Vidar Melby (BSc Hons) Nursing, Course Director, School of Health Sciences, University of Ulster, Cromore Road, Coleraine BT52 1SA. Tel: 01265 44141 - extension 4489.

Further Information
Further information on nursing in Northern Ireland can be obtained from:

The Recruitment Officer, National Board for Nursing for Northern Ireland, Centre House, 79 Chichester Street, Belfast BT1 4JE.

Wales

Diploma Courses
There are four institutions offering diploma courses in Wales. Details are given in Chapter 12. Applicants apply directly to the institution concerned. There are two intakes of students each year.

Degree Courses
Wales offers two degree programmes:

The three-year Bachelor of Nursing degree begins in autumn. It is offered in Adult, Mental Health Nursing and Children's Nursing at the University of Wales, Swansea. The Bangor campus of the University of Wales offers a degree programme in all four branches of nursing.

These three-year programmes are unique because students are eligible for the diploma bursary and not the student grant.

The four-year Bachelor of Nursing degree begins in autumn. It is offered in Adult, Mental Health Nursing and Children's Nursing at the University of Wales, Cardiff. North East Wales Institute of Higher

Education offers a four-year degree in Adult Nursing. Contract Plas Coch, Mold Road, Wrexham, tel: 01978 290666. Students on these programmes are not eligible for bursaries.

Applications should be made directly to the institution offering the course.

Further Information
Further information on nursing in Wales is available from:

Welsh National Board, 13th Floor, Pearl Assurance House, Greyfriars Road, Cardiff CF1 3AG.

Scotland

Diploma Students
The Centralised Applications to Nursing and Midwifery Clearing House (CATCH) is a central clearing system used throughout Scotland. It processes applicants from initial enquiry to placement. In September 1996 all colleges of nursing in Scotland were integrated into the university sector. Chapter 12 gives details of diploma courses. People wishing to apply for the autumn 1997 intakes should contact the National Board of Scotland (NBS) and obtain the *NBS 1997 CATCH Guide to Diploma Courses in Nursing & Midwifery*; address: NBS CATCH, PO Box 21, Edinburgh EH2 1NT.

Degree Students
Degree students apply via UCAS. The NBS provide a general information package for all Scottish nursing degree courses.

The Nursing Advisory Services, Scottish Health Service Centre, Crewe Road South, Edinburgh EH4 2LF.

Chapter 2
Midwifery

Midwifery is a profession quite distinct from nursing, although they share some common characteristics. Midwives have long been professionals in their own right, while nurses have continued to make great strides along the road to acceptance as professionals. Midwives do, however, encounter some difficulties in practising as professionals within a hospital setting as the medical profession tends to see pregnant women as 'sick' and not as healthy women undergoing normal changes within pregnancy. Nursing, despite recent initiatives in health promotion, is a therapeutic and curative profession. The majority of patients see nurses when they have health problems and expect a cure or an improvement in the disease.

Midwives have been nursing patients as people and not as diseases for many years. They work on health promotion and provide care and support, not only for the pregnant woman or the new mother but for the whole family. The role of the midwife is to help mothers to maintain good health during pregnancy and 90 per cent of pregnant women are healthy.

Working as a Midwife

Midwives are fully qualified to care for the pregnant woman throughout her pregnancy, to deliver babies and provide post-natal care for mothers. The midwife normally offers post-natal care for 28 days after delivery but this can be extended if a woman requires care arising from childbirth; in some cases, the midwife cares for a woman for many months, though not on a daily basis.

The work of the midwife is so much more than simply delivering babies. It involves offering support and health education to pregnant women; running parent-craft classes which help mothers-to-be to learn about baby care before the baby arrives; running ante-natal classes and explaining the normal changes of pregnancy. The midwife is able to offer advice on diet and exercise during pregnancy. She usually takes parents-to-be on tours of the hospital midwifery unit and explains the equipment and the usual routine in hospital.

Midwives run their own clinics and conduct interviews with newly pregnant women. They make detailed notes, normally referred to as histories, about the pregnant woman's general health. They monitor a mother's physical and psychological health throughout her pregnancy.

Pregnant women are usually also seen by doctors specially qualified in caring for pregnant women and by their own GP. There have been recommendations that midwives should be allowed more freedom in caring for women with normal pregnancies. The Changing Childbirth initiative announced by the Department of Health will lead to midwives taking the leading role in caring for many pregnant women.

Women who are pregnant for the first time, women who have had difficult pregnancies in the past, and women who are ill or having a difficult pregnancy have a high level of medical supervision.

Training

For many years, it was possible to train as a midwife without being a qualified Registered General Nurse (RGN). This was known as direct entry training and was very popular, especially with mature women who had children of their own and who wished to train as midwives when their children were a little older. Direct entry midwifery was discontinued some years ago but has now been reintroduced as part of the continuing development of midwifery education.

There are now two ways to become a midwife. The first is to take nurse training in the adult branch and then, after a period of experience as a staff nurse, to undertake a further period of training. The length of post-basic training is 18 months/or degree in Midwifery. This training allows the registered nurse to become a registered midwife (RM) and includes a diploma in higher education.

The second way is to undertake a pre-registration midwifery programme. This course lasts at least three years and on completion the student is entitled to be a registered midwife and has a diploma in higher education or a degree.

Applying for Direct Entry Training

The normal entry requirements are five GCE O levels or five GCSEs in grades A, B or C. One of these is normally required to be a science subject and another English. It is possible to take an entrance test if you do not have the required entry qualifications. Chapter 1 gives details of the DC test.

Degree Level Entry Programmes

There are now a number of courses leading to the qualification of registered midwife combined with a first degree. These courses are three or four years long and are very popular. They have long waiting lists and prospective students are advised to apply early.

The three-year courses offering an accelerated degree and midwifery qualification are very intensive. Candidates are carefully selected to ensure that they can manage this very condensed degree course.

Applying for Pre-registration Programmes
England
Applicants for degree programmes apply via the clearing house system, details of which are given in Chapter 1.

Applicants for pre-registration diploma programmes contact the institution. Details of pre-registration diploma and degree programmes for England are given in Chapter 12.

Northern Ireland
No pre-registration midwifery programmes are available in Northern Ireland at present. Midwifery education is developing all the time so applicants should check the current situation. The address for enquiries is given below.

Wales
Details of available pre-registration programmes can be obtained from:

Welsh National Board for Nursing and Midwifery, 13th Floor, Pearl Assurance House, Greyfriars Road, Cardiff CF1 3AG; tel: 01222 395535.

Scotland
Details of available pre-registration programmes are available from:

The Nursing Adviser, Scottish Health Service Centre, Crewe Road South, Edinburgh EH4 2LF. Tel: 0131 332 2335.

Applying for Post-registration Midwifery Programmes
The ENB or relevant national board provide a list of colleges which carry out post-registration midwifery training. At present the prospective student applies directly to the college concerned. There are plans to introduce a clearing system for post-registration midwifery training but no date has yet been set for this.

England
Chapter 12 gives details of institutions offering pre- and post-registration midwifery programmes.

Northern Ireland
Only one post-registration programme is available in Northern Ireland. There is one intake each year and competition is intense. Details can be obtained from:

Director of Midwifery Education, Northern Ireland College of Midwifery, Broadway Education Centre, Belfast BT12 6BB; tel: 0232 322565.

Pay and Conditions
Midwives have the same pay and conditions as registered nurses but registered midwives are usually paid on a minimum of an E grade (see Table 5, page 29). For the last year midwives have fought to have their

status as an independent practitioner recognised by F grade salary. To date this has been unsuccessful.

Working as a Qualified Midwife

After qualification the registered midwife gains general experience in a number of different areas, such as the labour ward, ante-natal clinic and special care baby unit. She may then decide to undertake further education or training or to seek promotion.

Working as a Community Midwife

There is currently no specialist training course for community midwives but midwives spend a substantial part of their training period working in the community. They must have two years' post-registration experience before applying for a post as a community midwife. Community midwives' posts are very popular and there is a lot of competition for them. They are usually advertised in professional journals and on the notice boards of local hospitals.

Community midwives work in a variety of settings; they are increasingly setting up their own clinics, some of these are in local hospitals and health centres where pregnant women from the local area receive antenatal care. Some midwives are based at GPs' surgeries and offer ante- and post-natal care while others offer post-natal care only. It seems that in future midwives will undertake routine ante- and post-natal care in clinics.

Community midwives also care for women in their own homes and in hospital. There are several schemes which allow a woman to choose where and how her child is born. The midwife can deliver a baby at home. She may also be involved in a scheme called a 'domino delivery', under which the mother is cared for by her community midwife who undertakes all routine ante-natal care, goes with the mother to hospital and delivers her baby. When mother and baby are comfortable after delivery the community midwife leaves them in the care of the hospital-based midwifery staff and returns, usually six hours later; if all is well the community midwife discharges mother and baby home. The midwife then meets the family at home and ensures that mother and baby are comfortable before leaving them. She visits once or twice daily as required for the first ten days and may continue to visit on a less frequent basis for 28 days after delivery. Community midwives always leave mothers a telephone contact number in case of an emergency.

Refresher Courses

It became a legal duty for midwives to undertake refresher courses under the Midwives Act 1936, and all practising midwives are obliged to undertake a one-week refresher course every five years. There are three ways in which midwives can fulfil these requirements:

1. A week's refresher course which can be residential or non-residential.
2. Seven accumulated approved study days over the five-year period.
3. A two-week clinical refresher course.

Midwives often choose the clinical refresher course if they are being asked to move to a different work area; for example a midwife who has worked for some years in a neo-natal intensive care unit (a unit where very sick and premature babies are cared for) might be asked to work on a labour ward and she would wish to update her skills before going to work in that area. There is also provision for individually designed refresher courses, though these are less usual.

Midwives who have undertaken some national board courses or the Advanced Diploma in Midwifery (ADM) can gain a five-year exemption from taking a refresher course.

Returning to Practice

Midwives who have not worked as midwives for a minimum of 12 weeks in the last five years or who have not notified their intention to practise must undertake a return to practice course. Midwives working abroad may have gained a wealth of experience and enhanced their skills in many ways but may, because of the different way that midwifery is organised in other countries, need to update certain aspects of their practice, such as ante- or post-natal care.

All refresher courses are individually tailored to the midwife and the length of the course is dependent on the length of time the midwife has not been practising. The content of the course will also vary.

Further Information
Further general information can be obtained from:

The Royal College of Midwives, 15 Mansfield Street, London W1M 0BE

Vikki Nix, Midwifery Adviser, Royal College of Nursing, 20 Cavendish Square, London W1M 0AB.

Further Education for Midwives

Further education for midwives follows the same pattern as that of nurses. Midwives can undertake diploma- and degree-level studies using the Credit Accumulation Transfer schemes (CATs) and Accreditation of Prior Experiential Learning (APEL) and Accreditation of Prior Learning (APL) procedures to build up a diploma or degree by undertaking a series of modules. This is explained in detail in Chapter 5 at page 42.

Post-registration Courses for Midwives
Courses run for midwives are more flexible than those run for nurses. Many are run by colleges of midwifery on a 'one off' basis and no specific booklets are available. These courses are, however, of the highest

academic content, are validated by universities, and carry points at either level two (diploma level) or level three (degree level).

Midwives may also spend time working with clinical nurse specialists in order to gain skills which relate to their practice. For example, a midwife may spend a day with a clinical nurse specialist updating and improving her knowledge of diabetes so that she can provide the highest quality of care for pregnant women who suffer from diabetes.

ENB Courses for Midwives

There are a number of national board courses open to nurses and midwives. The following are some examples:

ENB 280. This course enables the midwife to care for people who are HIV positive or who are suffering from AIDS. Midwives must have undertaken ENB 934 before they are allowed to do this course. ENB 934 is a 10-day course which serves as an introduction to this subject.

ENB 405. This course enables the midwife to care for babies requiring special and intensive care. It lasts 24 weeks and equips the midwife with the skills required to work in a special care baby unit with very ill and premature babies.

ENB 870. This course enables midwives to have a greater understanding and appreciation of research. It lasts 40 days and is usually spread over a year or less. It does not qualify the midwife to undertake research.

ENB A05. This course enables midwives to provide high levels of care for diabetic mothers in their care in either a hospital or a community setting. The course lasts 60 days and is spread over a year or less. It can be arranged in study blocks or on a day-release basis.

ENB A06. This course enables midwives to counsel (to listen to an individual's problems and allow each person to work out the solutions to her problems within a supportive setting). The course lasts 46 days and is spread out over 36 weeks.

ENB A07. This course enables the midwife to care for patients with genito-urinary infections and related disorders. The course lasts 60 days; 20 days are theory and the remainder is practical experience. This course is intended for nurses working within a department of genito-urinary medicine.

ENB A08. This course allows experienced family planning nurses to develop advanced skills and undertake total care of family planning clients.

Chapter 3
Working as a Qualified Nurse

Working as a qualified nurse is very different from working as a student nurse. It is usual for a newly qualified nurse to apply for her first post as a staff nurse within the hospital where she gained clinical experience. She normally works in a non-specialised area such as a medical or surgical ward if trained to care for adults and would not normally work in a specialist area such as an intensive care unit until she has some general experience. There is an acute shortage of registered nurses at the time of writing (August 1996), and they can choose the area they wish to work in.

The first six months after qualification is a very important period in the newly qualified nurse's career. It offers her the opportunity to combine theory with practice. During this time diploma and degree qualified nurses must learn to reconcile the freedoms which they experienced as a student with the responsibilities of the qualified nurse dealing with a varied workload. They must master a range of clinical skills and begin to assume responsibility for planning care.

The United Kingdom Central Council (UKCC)) have issued guidelines which state that every newly qualified nurse must have a period of supervised practice and that she should have a more experienced qualified nurse as a mentor, to guide her and assist her in making the transition from student to qualified and experienced practitioner. The period of mentorship is normally six months, though some newly qualified staff may require an extended period of support, perhaps because they work part time.

The newly qualified nurse is normally offered a D grade post (see page 28 for an explanation of grades) and does not normally take charge of the ward in which she is working. She may be expected to take charge in an emergency but can always call on more senior nurses who can provide help and support. The newly qualified nurse will often work in a unit where primary nursing is undertaken and will care for a group of patients. She will usually work with a more experienced registered nurse who will be an E grade nurse or above. The newly qualified nurse will be able to work with her colleagues and patients to provide quality care.

Primary Nursing

One of the greatest changes in nursing over the last few years has been the abandonment of task allocation. Task allocation was a method of dividing up the workload according to the level of skill each nurse had attained. It developed during the time when all student nurses, including very junior students, were part of the workforce. It was possible, using task allocation, to care for large numbers of patients with small numbers of qualified staff and large numbers of junior students. Nurses were allocated specific tasks: for example, one nurse was asked to take all the patients' temperatures or to give out all the patient medications.

The primary nursing system means that each patient has a 'primary' or named nurse responsible for his care during his stay on a hospital ward. The primary nurse is assisted by one or more associate nurses who care for the patient in the absence of the primary nurse.

Newly qualified staff nurses usually work as associate nurses; when they have developed sufficient expertise they move on to become primary nurses. Senior nurses with managerial commitments sometimes work as primary nurses for part of their working week to maintain their clinical skills.

The primary nurse builds a relationship with her patient and involves him in planning care which is appropriate to him as an individual. This system is now acknowledged to provide higher quality care than task-orientated nursing. It allows patient and nurse to form a closer relationship than before; this reduces patient anxiety and provides a less fragmented type of care. Nursing research has shown that patients who are relaxed and have a trusting relationship with nursing staff recover more rapidly, suffer less pain and have a reduced rate of complications. It is now government policy that all patients should have a primary nurse, though at present primary nursing has only been introduced to about 20 per cent of units.

Under the old system of patient allocation, the nurse would care for a group of patients for a few days and then be switched to another group. This was to prevent nurses becoming too involved with their patients. It was thought that if nurses became too close to patients and did not maintain a distant relationship, they would be unable to care for patients without unbearable stress levels. It was also felt that it would lead patients to become over-dependent on particular nurses and that this would hinder their recovery. These concerns may have been justified under the traditional system of nurse training where very young junior student nurses were part of the workforce. The staff nurse now either works as a primary or associate nurse, depending on her level of experience.

Primary nursing helps nurses to continue to improve their skills as nurses, as the patient queries and paperwork are now shared more equally between all trained nurses.

Induction Courses

When a newly qualified nurse begins work, she is usually asked to attend an induction course. These courses are carried on within the hospital or group of hospitals where she works and there are usually several staff on each course. The courses explain practical points such as where certain departments are located and what services they provide. They also cover basic aspects of the management role of the staff nurse, current issues in nursing, hospital policy on many issues and some guidance on the teaching and support of more junior colleagues. There is usually the opportunity for the nurse to have parts of the course tailored to her individual needs. The courses vary in length, as a nurse who had spent time in the hospital as part of her training would require less information than a nurse new to the hospital.

Teaching and Supporting Colleagues

Many newly qualified nurses feel unsure and uncertain when they first qualify; this is entirely normal and the newly qualified nurse will gain confidence as she gains experience. It can come as a shock to realise that student nurses are even more unsure than she is. Teaching student nurses is an important part of the staff nurse's role. Many students will turn to her for help and guidance in clinical care. Other students may be feeling homesick or unsure of their role. Usually, students find newly registered nurses (RNs) very approachable and ask them for support and guidance.

The qualified nurse has a duty to keep up to date, learn new skills and develop her areas of expertise. Her actions will be quietly watched by student nurses who will often model themselves on her.

The qualified nurse should always make time in a busy day to explain procedures and treatments to student nurses and should check that they understand the reasons for different types of patient care. She will often set aside time for formal teaching sessions where she will explain the types of diseases which patients on her ward suffer and how nursing care is related to the patient's condition, expectations and feelings.

Many hospitals are now training health care assistants and in future staff nurses will become more involved in teaching HCAs and encouraging them to become skilled at caring for patients under their supervision. There are a number of courses which will help the staff nurse to develop her skills as a teacher. The ENB course 998 Teaching and Assessing in Clinical Practice is essential for all staff working with students. This course contributes to the staff nurse's own professional development (see Chapter 4) and is usually run within hospitals.

Career Development

About six months after qualification, the staff nurse remembers her lack of confidence with astonishment and comments that she really began to learn far more about nursing after obtaining her basic qualification.

At this point, the staff nurse has more idea about the area in which she wishes to work. She may decide to continue to work in hospital; she may wish to undertake a further period of training; or she may wish to work in the community. Post-registration training is discussed in detail in Chapter 4; community nursing is discussed in Chapter 7; midwifery training in Chapter 2.

The staff nurse usually goes to work in another area after six months' experience. If she has worked on a surgical unit and she wishes to remain in hospital work, she may move to a medical ward (or vice versa).

Many hospitals run staff nurse development courses. They aim to build on the staff nurse's new skills and to allow her to become a more effective nurse.

It is often possible for a nurse to gain promotion to an E grade staff nurse after a year's experience. The rate of promotion is dependent not only on the nurse demonstrating her skills but also on posts becoming available. Conditions in the NHS, at the time of writing, may mean that nurses will have to wait longer to attain promotion.

Career Options

Staff nurses can pursue any career that they wish in nursing: all options are open to them. They may choose to become ward sisters, or ward managers as they are becoming known. The ward manager has a different role from that of the traditional ward sister and may work 9-5 Monday to Friday. Her role is increasingly becoming more managerial, as the title suggests, and often today's ward manager has little time for 'hands on' nursing. She may be involved in supervising a medical unit in the absence of the senior nurse in charge of the unit. She may be needed to supervise and support less experienced nurses on her own ward or throughout the unit. She has an important part to play in teaching and supporting student nurses and staff nurses.

Staff nurses usually work for a minimum of two years before obtaining a ward manager's post. The staff nurse progresses through posts as a D and E grade nurse and usually obtains an F grade post which is classified in some hospitals as a senior staff nurse and in others as a junior sister. The ward manager has continuing 24-hour responsibility for the care delivered on her ward. This care is given by other nurses when she is off duty but the responsibility is hers (although, of course, each individual nurse is accountable for her own actions).

There is more emphasis on specialisation than ever before and, if a nurse is interested in working in a particular specialist area, she is encouraged to undertake a post-registration course in that speciality. Nurses working without a post-registration course in a specialist area would find it very difficult to obtain promotion. In a specialist area, nurses without post-registration training in that area would find it difficult to obtain any post above an E grade.

A ward sister working in a field such as neurosurgery or cardiology in a large teaching hospital would be expected to have completed a post-registration course and to have obtained clinical experience after the course. Further details of post-registration courses are outlined in Chapter 4.

Working Conditions

Staff nurses normally work full time and the full working week is 37.5 hours (excluding meal breaks). There are opportunities for staff nurses to work part time or to work hours which are suitable for those nurses with domestic responsibilities. This is explained in greater detail in Chapter 6. Staff nurses working in hospital normally work shifts, although nurses working in certain areas such as outpatients departments or day surgery units may not be required to work shifts. Nurses working in outpatient departments normally work only Monday to Friday and do not usually have to work in the evening.

Standby Duties

Nurses working in areas such as operating theatres may work only morning and evening shifts but have periods 'on standby'. This means that they are not working but are expected to leave a telephone number where they can be contacted. They may be called to the hospital to work, for example, at night or at the weekend, when normally only emergency operations are carried out. There are always some theatre nurses in the hospital so that they can quickly prepare for emergency theatre cases, but if there are a number of urgent theatre cases the standby nurses will come in to assist.

Nurses on standby duties are paid an allowance for each standby shift. They receive full pay if they are called into the hospital. The standby allowances (since April 1995) are:

Overnight: £9.70
Weekend: £13.20
Public holiday: £16.85.

On-call Duties

At times the senior nurse provides support for her colleagues on duty and can be contacted via a long range radio-pager which she carries with her when she is off duty. An 'on-call' allowance is paid for each shift the nurse is on call. The on-call allowances (since April 1995) are:

Overnight: £4.85
Weekend: £7.25
Public holiday: £9.70.

Shift Patterns

Internal rotation, a system which means that all nurses work a series of day and night shifts, is becoming very popular. This means that nurses who once worked only at night are now expected to work a full range of shifts, and nurses who worked only day shifts now also have to work on night duty. The need to provide trained nursing staff every hour of every day has led many hospitals to ask nurses to work an 85-hour fortnight instead of a 37.5-hour week. This means that nurses may work a long stretch, perhaps eight days in a row, but then have four days off together. This can be tiring.

Qualified nurses are expected to work a proportion of bank holidays but this is normally shared out fairly, with each nurse working part of a holiday period. A nurse who works over Christmas, for example, usually has time off over the New Year (or vice versa). Qualified nurses have five weeks' paid holiday per year and a further ten days of bank holiday entitlement because they are often working over holiday periods. Nurses are paid extra for working unsocial hours; for example, at weekends, on bank holidays, after 6 pm during the week and on night duty.

Special Allowances
Nurses who work in specialist areas such as psychiatry are paid a small extra allowance; this is called a 'lead allowance' and is referred to as a 'geriatric' or 'psychiatric' lead. The annual allowance rates (April 1996) are:

Psychiatric lead: £425
Geriatric lead: £165
Regional secure unit: £995.

London Allowances
Nurses who work in London are paid an extra allowance which helps to meet the extra expense of living there where rents, for example, are more expensive than in other areas. The London allowance is a fixed sum plus a percentage of salary up to a maximum amount. The rates (April 1996) are:

Inner London: £1,910 plus 5 per cent of salary up to a maximum of £750
Outer London: £1,360 plus 5 per cent of salary up to a maximum of £750
Fringe zone: £245 plus 2.5 per cent of salary up to a maximum of £375.

Clinical Grading
Clinical grading was introduced in 1988 to reward nurses more fairly for the work they carried out. It was intended to recognise and reward the clinical skills of nurses who chose to continue to work in clinical areas.

The minimum salary scale for enrolled nurses is grade C. A newly qualified enrolled nurse or one who carries out basic duties will be paid on grade C. An enrolled nurse with specialist skills, such as intensive care qualifications, will normally be paid at a grade which reflects the increased responsibilities of her post. This may be a grade D, E or even F.

The minimum salary scale for a registered nurse is grade D. Newly qualified nurses normally spend a period of time working as a D grade prior to obtaining an E grade post. Nurses who are employed as a D grade normally work under the supervision of more senior nurses and do not normally take charge of a ward or department. Nurses employed as E grade normally supervise more junior nurses and regularly take charge. The majority of staff with specialist skills (for example, an ENB qualification in neurosurgical and neurological nursing or a midwifery qualification) are employed on a minimum of E grade.

The minimum salary scale for a sister (or charge nurse) is F grade, though sisters who have continuing 24-hour responsibility for a ward or department are graded G. The minimum salary scale for qualified district nursing sisters and health visitors is grade G.

Nurses with specialist skills and expertise are called clinical specialists and are normally paid at H or I grade. The grade is dependent on the responsibilities of the post. Nurse managers are also normally paid at H or I grade.

Current Salary Scales for Qualified Nurses

In April 1996 the body responsible for setting NHS nurses' salary scales recommended a 2 per cent national increase and, as in 1995, attempted to introduce local pay. Under local pay arrangements each hospital or community unit (known as an NHS Trust) could 'top up' the national award with a further award of up to 4.5 per cent. The aim was to enable trusts to tailor pay rates to the local labour market. This move has been bitterly opposed by the nursing unions. At the time of writing (August 1996) the April 1996 pay rise has not been settled. Table 5 gives details of nurses' salary scales with a 2 per cent increase and a 6.5 per cent increase.

Table 5. Nurses' Recommended Salary Scales

Grade	Salary range with 2% increase	Salary range with 6.5% increase	Comments
C	£10,374-£12,292	£10,832-£12,834	This grade is only used for some Enrolled Nurses.
D	£11,895-£13,605	£12,412-£14,199	This grade is the starting grade for newly Registered Nurses.
E	£13,606-£15,759	£14,206-£16,454	This is the grade for experienced Registered Nurses who take charge of a ward or unit on a regular basis.
F	£15,097-£18,491	£15,763-£19,306	This grade is described as 'Senior Staff Nurse' or 'Junior Sister'.
G	£17,979-£20,592	£18,582-£21,500	This grade is used for sister/charge nurses or ward managers who have 24-hour responsibility for a unit.
H	£19,888-£22,746	£20,765-£23,750	Nurses at this level provide clinical advice and support to nurses and other professionals.
I	£22,021-£24,952	£22,922-£26,053	This is the most senior clinical grade. I grade posts are becoming rare.

Compulsory Further Education for Qualified Nurses

The introduction to the Post-Registration Education and Practice (PREP) legislation in 1995 compels nurses to continue their professional development in order to remain on the professional register. Nurses are unable to practise unless they can demonstrate that they have undertaken post-registration education within the previous three years. This is discussed in detail in Chapter 5.

Chapter 4
Post-Registration Qualifications

Nurses* are now expected to continue their education and to improve and develop their practice throughout their working lives. There are more opportunities than ever before for nurses to gain new skills.

All courses which can be recorded by the qualified nurse as part of her qualifications must be approved by the relevant National Boards for Nursing, Midwifery and Health Visiting. The NBs approve courses run by individual colleges and ensures that all colleges provide high quality courses by sending inspectors to colleges and carrying out checks on courses.

There are courses which lead to an additional qualification on the nursing register. The registered nurse can, for example, also become a registered sick children's nurse, or a midwife. There are courses which allow a nurse to develop a specific area of expertise; for example, the ability to care for patients needing intensive care, cardiac care, or care and rehabilitation in old age.

In the past, courses which were less than two months long were treated differently from courses over two months. This distinction has now ended and all courses are now called post-registration awards instead of certificates, statements of attendance or certificates of competence. This change should end the confusion that nurses and members of the public faced when so many different types of award were available.

There has been an expansion in the number of courses available at diploma level and the introduction of Credit Accumulation Transfer schemes (CATs) means that nurses can acquire points and build towards a diploma or degree qualification. The introduction of the awards by credit accumulation for professional development (APEL and APL), explained on page 43, has enabled nurses to build upon their previous experience and courses undertaken to acquire a degree in nursing. The introduction of the ENB Higher Award is intended to link theory with practice. Open learning has allowed nurses to update and also to convert from enrolled nurse status to registered nurse status in their own time and at their own pace.

---

* The term 'nurses' is used for convenience and embraces nurses, midwives and health visitors.

Sick Children's Nursing

An RGN (Registered General Nurse) wishing to become a sick children's nurse normally undertakes a 53-week course. An RGN who has a great deal of experience in caring for sick children can apply for a shortened training course which is 26 weeks long. Only one centre is currently offering this course, but check the situation with the relevant national board, as the situation may change. Addresses are given in Chapter 1.

Registered midwives who have undertaken the direct entry midwifery training (outlined in Chapter 2) can undertake a 104-week course. RNMH and RMN (Registered Mental Nurse) qualified nurses can undertake a 73-week training course. Enrolled nurses who have been working with sick or handicapped children for a minimum of one year after qualifying can undertake a 97-week course. Enrolled nurses with the ENB 426 course in paediatric nursing and a minimum of one year's experience in caring for sick children can apply for a shortened 52-week course.

Specialist Courses for Sick Children's Nurses

There are a number of courses available which are specially designed for children's nurses to improve their skills and work in specialist areas. These include:

Post-registration clinical course in paediatric renal nursing: course no 147
Post-registration course in paediatric medical and surgical cardiothoracic nursing: course no 160
Post-registration course in paediatric oncological nursing: course no 240
Post-registration course in special and intensive nursing care of the newborn: course no 405
Post-registration course intensive care nursing of the children: course no 415

These courses, all 24 weeks long, are just some of those available for the RSCN and a number of shorter courses are available. It is recommended that all RSCNs have at least six months' experience in caring for children after qualification before beginning a specialist course. There are often long waiting lists for courses so you may wish to apply immediately you qualify to avoid a long wait for a place.

An ever increasing number of posts are available for RSCNs within community nursing and the RSCN's expertise could lead her to consider a career as a district nurse or a health visitor. Further details about working as a community nurse are given in Chapter 7.

Further Information

The national boards produce booklets giving details of career opportunities and first level courses in children's nursing. Addresses of the national boards are given in Chapter 1.

It is important to note that, although a college has approval to run a particular course, this does not mean that the course is being run at present. It is important to check with the college if the course is still being offered. Some courses and training colleges have very long waiting lists. The more prestigious and popular colleges tend to demand higher academic qualifications and competition for places at world-famous hospitals such as the Hospital for Sick Children at Great Ormond Street is intense. Nurses from all over the UK and abroad apply for training places at the colleges which are attached to such hospitals. It is advisable to apply early and to choose a few places where you would like to train. This avoids disappointment.

Mental Health Nursing

Many registered general nurses undertake this training after qualification as they are eager to nurse the whole patient and not just the body. The importance of caring for the patient as a human being and not simply as a person suffering from a physical disease is now emphasised in all branches of nursing. Many nurses feel that a qualification in mental health nursing (which nurses usually refer to as psychiatric nursing) will equip them with the necessary skills to help patients become mentally and physically healthy.

The course, normally 73 weeks long, is very different from general nursing. The atmosphere in psychiatric nursing units is more relaxed than in general units and this can come as something of a shock to RGNs, particularly those who trained under the traditional system some years ago.

The course leads to registration on Part 3 (Mental Health) of the professional register.

Application
Nurses can obtain information about courses from the relevant national board. Alternatively, you can obtain details of colleges from Chapter 12 and check if they offer the course. The lists from the national boards are updated periodically and can be out of date.

Specialist Courses for Registered Mental Nurses
There is an enormous range of post-registration courses available to registered nurses with mental health qualifications. Some courses enable the nurse to nurse patients suffering from alcohol and drug dependency. There are courses in psychotherapy, rehabilitation, behaviour modification, nursing within secure units, etc, as well as courses in community nursing (see Chapter 7 for details). A few of the courses available are listed below. Further details are available from the national boards.

Drug dependency nursing for nurses, midwives or health visitors ENB 616
Drug & alcohol dependency nursing for nurses, midwives or health visitors ENB 612
Alcohol dependency nursing for nurses, midwives or health visitors ENB 612
Adult Behavioural Psychotherapy for Nurses on Parts 3 (RMN) or 4 (EN(M)) of the register: short-term therapy ENB 650

Rehabilitation in residential settings for nurses on Parts 3 (RMN) or 4 (EN(M)) of the professional register ENB 665

Nursing within controlled environments for nurses on Parts 3 (RMN), 4 (EN(M)), 5 (RNMH) or 6 (EN(MH)) of the professional register ENB 770

Nursing care of mentally ill people in the community on Part 3 (RMN) of the professional register ENB 811: this is a 36-week course which educates and trains RMN nurses to care for and support patients living in the community. It emphasises the need to provide help and support not only to the patient but also to the patient's close family.

Learning Disability Nursing

There are a number of colleges which offer a 53-week training course which enables nurses qualified in adult nursing to gain a qualification in learning disability nursing. There are no plans to discontinue learning disability nursing although the focus of this qualification has changed from hospital to community.

Specialist Courses for Registered Nurses Specialising in Learning Disabilities

In recent years, the focus of caring for people with learning disabilities has changed from institutions to care in the community (see Chapter 1). The move from the old institutions into the community setting is almost complete. Now most people with learning disabilities live in small houses and are enabled to become as independent as possible. A post-registration course (ENB 807) is available to enable the nurse to acquire the skills necessary to support her patients in the community. The course is 36 weeks long and incorporates a period of supervised practice.

There are other courses available to the learning disability nurse. For example:

Diploma in Social Learning Theory ENB A02
Developing services for people with a learning disability ENB A04
Counselling for nurses, midwives and health visitors.

Further details of post-registration courses are available from the ENB or the relevant national board.

Adult Nursing

Nurses who have trained in a specialist branch of nursing may wish to acquire a qualification in adult nursing. Nurses with dual qualifications, for example adult and mental health nursing, are much in demand and have excellent promotion prospects.

Adult Nurse training for Mental Health Nurses

This course lasts 73 weeks and is available at only a few centres at present. Nurses wishing to undertake this course are advised to apply early as waiting lists can be long. Details of courses are available from the relevant national boards.

Adult Nurse Training for Learning Disability Nurses
Although many nurses registered as learning disability nurses wish to obtain qualifications in adult nursing, there are no colleges with approval to run this course at present. When available the course lasts 73 weeks. In view of the demand it may be offered in the near future. Check with the relevant national board. Currently, many learning disability nurses have to complete a full diploma programme to obtain an adult nursing qualification.

Registered General Nurse Courses for Registered Sick Children's Nurses
There are no colleges offering this course at present. When available the course lasts 73 weeks. It may be offered in the future and advice can be obtained from the ENB or the relevant national board.

Enrolled Nursing

Enrolled nursing was introduced in 1943 when it was clear that registered nurses were unable to provide care for very dependent patients with only student nurses to assist them. They were then called state enrolled assistant nurses. The 'assistant' was dropped from the title in the early 1960s and post-basic training was developed for enrolled nurses.

Enrolled nurses worked in every area of nursing: psychiatric hospitals, medical and surgical wards and the community all benefited from their contribution. They provided a stable workforce and were often concentrated in areas in which it was difficult to recruit registered nurses, such as elderly care and learning disability.

Enrolled nurse training was discontinued in 1992 and some 100,000 enrolled nurses remain in practice. There are some in the nursing profession who feel that the discontinuance of enrolled nurse training is a mistake which the profession will regret in future years. The introduction of NVQ training is the reintroduction of enrolled nursing under another guise.

The Future for Enrolled Nurses
Many enrolled nurses feel very anxious about their future. It is more difficult for enrolled nurses to find jobs than before and the changes in the health service have led to a shortage of jobs. Many employers are keen to employ newly qualified RGNs in preference to enrolled nurses.

The enrolled nurse now has to face one of two options. She can remain as an enrolled nurse or she can convert to become a registered nurse in either general nursing, learning disability, mental illness or children's nursing. The type of conversion she undertakes depends on her present qualification.

Conversion Courses for Enrolled Nurses
These courses lead to first level registration and are slowly beginning to recognise the enrolled nurse's previous experience and training. Two different types of course are available to the enrolled nurse. The first is a place on a conversion course at a hospital. The second is to undertake an open learning course to obtain first level registration.

Hospital-based Conversion Courses
These courses are available in many colleges of nursing. Two types of course are available for enrolled nurses trained in general nursing: a full-time course which is normally 52 weeks long, and part-time courses which are 104 weeks long and may appeal to enrolled nurses who choose to work part-time because of domestic commitments.

Competition for places on these courses is keen and nurses usually have to have one year of post-registration experience appropriate to the type of course, such as adult nursing for RGN, before being considered.

Enrolled nurse training did not include maternity care, care of the newborn, community care and mental health care and these areas are included on the course.

Case Study
A nurse who recently completed the conversion course shares her experience.

Sally Page completed her enrolled nurse training in 1986 at the Westminster Hospital. She began her search for a place on a conversion course almost immediately after qualifying.

> I felt frustrated with my role as an enrolled nurse at times. I was good enough to run the ward one day and the next I was the most junior qualified nurse. I wanted to progress in my career and felt that the opportunities to do so were limited because I was an enrolled nurse.
>
> I found it very difficult to obtain a place on a conversion course. I wrote to hospitals all over the country. A lot of them only considered nurses who were working in their hospital; others had a two- or three-year waiting list. I was working at St George's Hospital in London when they started offering the conversion course in 1989. It was difficult to obtain a place because I did not have any post-registration qualifications but I persevered and took GCSEs to show the college that I was capable of studying. I obtained a place on it in 1991 and did the course.
>
> There were 20 nurses on the course and we all found it hard work but very worthwhile.

Sally now has a senior post in elderly care and plans to continue her education by doing short courses and working towards her diploma in nursing.

Applying for a Conversion Course
The national boards produce a list of institutions approved to run conversion courses and you can obtain this by writing to them. You can then enquire about the availability of courses and apply to your selected training colleges.

Open Learning Conversion Courses for Enrolled Nurses

The journal *Nursing Times* pioneered open learning as a means of allowing enrolled nurses to convert to first level registration while continuing to work as normal. In the last five years 10,000 enrolled nurses have joined the programme and 98 per cent have successfully completed the course.

Programme Content

The programme consists of three integrated units and three specialist modules. The integrated units are:

1. Professional development
2. Management
3. Research

General trained enrolled nurses are required to gain experience in specialist areas to fulfil European Community Directives. The aim is to enable nurses to gain insight – not expertise – within these areas. Table 6 gives details of the specialist areas and the minimum number of hours required. These hours include the time taken to complete the workbooks supplied as part of the course.

Table 6

Specialist Area	Hours Required
Care of the mother and newborn	150 hours
Mental health and learning disability	150 hours
Community healthcare	60 hours

Entry Requirements

Working hours. The UKCC and national boards require students to be in full- or part-time employment. Students who work less than 20 hours each week may need to increase their working hours to fulfil course requirements.

Experience. The equivalent of at least one-year's full-time post-registration experience.

Support. The written support of your manager.

The programme must be able to link you to a centre offering the conversion programme.

Method of Assessment

Students are assessed continually throughout the programme. The course includes a formal written component that is timed and invigilated.

Costs

Students purchase learning materials from *Nursing Times*, costing £540.
  Students pay for tutor support, workshops, library facilities and clinical placements. These costs vary from centre to centre.

The total cost (including learning materials) ranges from £2,000 to £3,000.

Some employers are willing to fund the costs of conversion courses fully; others contribute part while some make no contribution at all.

There are many scholarships and grants available to nurses unable to obtain employer sponsorship.

The library of the RCN has a number of books and pamphlets which give details of sources of funding for further education for nurses. The address is:

Royal College of Nursing, 20 Cavendish Square, London W1M 0EB. Tel: 0171 407 3333.

Students who pay all or part of their own fees are entitled to tax relief.

Further Information
Nursing Times Open Learning, Ref: Conversion, Porters South, Crinan Street, London N1 9SQ. Enquiries Hotline: 0171 843 4850.

Case Study
Cecilia Odigie is 45 and the mother of four children. She trained in the West Midlands in 1971. Cecilia attends the nursing faculty of Kingston University once a fortnight for tutorial support. She hopes to complete her course in summer 1997.

> I decided to begin the course because I felt that I had no future as an enrolled nurse. I was terrified when I began. It was so long since I had studied. I wondered if I had the ability to become a first level nurse. Now that I've been studying for a year I realise that my fears were unfounded. I'm really glad that I've taken the plunge. I've been lucky my employer met the costs of the programme and gave me study leave. I would advise other ENs considering conversion to go for it.

Post-basic Courses for Enrolled Nurses
Many enrolled nurses hold post-enrolment qualifications and these can be converted to first level qualifications if they wish. It may be possible to do this as part of a conversion course or to do a short period of adaptation to convert the course to a first level registered course. You should enquire at your local college of nursing or the college where you intend to undertake a conversion course about this.

There are a few courses still available for enrolled nurses who decide not to convert and they include:

Short course in special and intensive care of the newborn ENB 904
Short course on the care and rehabilitation of physically disabled people ENB 913
Short course on the principles of intensive care ENB 918
The continuing care of the dying patient and the family ENB 931.

There are also a number of longer courses which are usually 24 weeks long. They enable the enrolled nurse to care for patients in specialist areas, and include:

General intensive care nursing ENB 115
Coronary care nursing ENB 130
Renal and urological nursing ENB 140
Nursing elderly people ENB 298
Ophthalmic nursing ENB 348.

The Future for Enrolled Nurses
Although the UKCC have stated that enrolled nurses still have a role within the NHS, younger nurses, under the age of 40, are undertaking conversion in increasing numbers. Older nurses planning to retire within the next few years tend not to convert.

Chapter 5
Continuing Education for Registered Nurses

Great changes in nurse education have occurred in recent years. The introduction of diploma-level basic training, moving nurse training into higher education, and compulsory post-registration education have all affected nursing.

In April 1995 legislation compelling nurses to continue their professional development in order to continue practising was introduced: Post-Registration Education and Practice or PREP. No government funding has been made available for it. Nurses can apply to their employers for funding and study leave. Employers, however, are not obliged to fund PREP. The situation varies from area to area. Recent research suggests that about half of all registered nurses fund their own professional development.

UKCC Areas of Relevance

Recently the UKCC issued guidance on what it will view as continuing education. This is defined as 'areas of relevance' which are listed below:

Reducing risk:
- health promotion identification
- protection of individuals
- risk reduction
- screening
- heightening of awareness

Care enhancement:
- developments in clinical practice and treatment
- new techniques and approaches to care
- standard-setting
- empowering consumers

Patient, client and colleague support:
- counselling techniques
- leadership for professional practice
- supervision of clinical practice

Practice development:
- external visits
- exchange arrangements

- personal research/study
- briefing on health and professional policy changes
- service audit

Educational development:
- external visits
- exchange arrangements
- personal study/research
- educational audit
- teaching and learning skills.

This list shows that nurses can use a whole range of activities to demonstrate their continued professional development and fitness to practise. The activities a nurse undertakes to maintain and develop her practice will be recorded in a 'personal profile'. The UKCC requires nurses to maintain a professional profile. The nurse can either make up her own profile or purchase a professionally prepared profile to complete.

Your PREP Requirements

The UKCC emphasise that you can fulfil your PREP obligations in a number of ways:

- Attending study days, courses and conferences.
- Visiting other units and places relevant to your work.
- Spending time in a library reading and researching.

Nurses are required to record the following information in their personal profiles:

- Relevance of the activity
- Your aims in undertaking the activity
- Assessment of the outcome
- Time spent on each event
- Your working hours over a three-year period.

Further Information

You can write to the UKCC or telephone their PREP Hotline:

UKCC, 23 Portland Place, London W1N 4JJ. Tel: 0171 333 6692.

Ways of Meeting PREP Requirements

You can attend study days, conferences and seminars. Study days offer you:

- The chance to meet other nurses working in the same field.
- Opportunities to discuss topics and gain ideas.
- The opportunity to ask the speakers questions and to clarify points.
- Opportunities to view any exhibitions, to see new products and to obtain samples.

However, study days can be expensive. They can involve travelling and sometimes an overnight stay. Some nurses, especially those with dependants, can find it difficult to attend study days. Finding out as much as possible about the study day, the topics and the speakers will help you choose relevant study days.

Open learning assessments printed in *Nursing Standard* and *Nursing Times* are another option. Open learning offers:

- Easy access, important to nurses working shifts and those with family commitments.
- The option to study when it suits you.

Also:

- Study leave is not required.
- Managers are more likely to fund open learning because it is less expensive.

However, busy nurses have to find time to study after a week's work – this requires self-discipline. There is less opportunity to meet other nurses and gain ideas.

Nursing Standard Open Learning

The Royal College of Nursing run an open learning programme called Nursing Update. Programmes in the series are broadcast on BBC television during the night. The nurse can videotape the programme and work through open learning material which is published in *Nursing Standard*. They then complete multiple choice assessments and send them off to be marked. Each assessment costs £10 to RCN members.

Further Information
Nursing Standard, 17-19 Peterborough Road, Harrow, Middlesex HA1 2AX. Tel: 0181 423 1066.

Nursing Times Open Learning

Nursing Times also publish open learning material for professional development. Normally, four units are published on a particular theme. Nurses are advised to retain the units in their portfolios. Those who complete a multiple choice assessment gain a certificate giving details of the study hours attained. The cost of an assessment is £8.

Further Details
Nursing Times, Porter's South, Crinan Street, London N1 9XW. Tel: 0171 843 4600.

These open learning programmes are acceptable as evidence of further study as required by PREP.

Diploma and Degree Studies

To gain a diploma you must have 120 CATs (Credit Accumulated Transfer scheme) points at levels one and two. To gain a degree you need 120 level two and 120 level three CATs points. Traditionally trained registered nurses with two years' experience are normally credited with

120 level one CATs points. To obtain a diploma they must obtain 120 CATs at level two. A further 120 CATs at level three are required for an honours degree.

Diploma-level students gain 120 level two CATs on qualification. They must obtain a further 120 level three CATs to obtain an honours degree.

CATs Points and Continuing Education

Credit Accumulation Transfer schemes (CATs) were established by the Council for National Academic Awards. They have been adopted by universities and colleges of higher education. CATs give a points rating to courses and modules. The points rating indicates the level of study and the estimated amount of time demanded by the course. CATs points can be transferred from one system to another.

Making Your Learning Count

Credits for Prior Learning

The APL (Accreditation of Prior Learning) scheme enables nurses to be awarded academic credit for previous learning. Courses such as ENB courses taken some years ago may be retrospectively assessed as having a certain number of CATs points. ENB courses usually carry CATs points at level two (diploma level). If you plan to begin studying for a diploma check on the current CATs rating of any courses you have completed. Usually, you can claim credit for previous studies. You may have to write an essay which will allow that work to be credited.

The APEL (Accreditation of Prior Experiential Learning) scheme enables experienced nurses to claim academic credit for work that has not been formally assessed. This involves collecting details and evidence of your work and applying for academic credit. You may, for example, have produced a teaching package for student nurses, or a discharge information pack for patients or introduced any number of changes within your workplace that benefited patients. The credits gained in this way can be used with further study to gain a diploma or a degree in nursing.

For further information about APL and APEL contact the college you have decided to study at.

Diploma-level Studies

College-based Studies

Diploma-level courses are run by colleges of nurse education and some colleges of further education. They are advertised in the nursing press. Your chosen college of nurse education can supply details of available courses. Addresses are given in Chapter 12. These courses usually involve day release over one or two years (depending on how many CATs points you have) for the academic term of 36 weeks. Modular courses offer more flexibility if your circumstances change. Many courses combine diploma-level studies with a national board qualification. The cost of diploma courses varies from one institution to another. The total cost is around £2,000-£3,000. In addition, you will also have to purchase books and pay for travelling.

Open Learning Options
Three institutions have been in the forefront of open learning at diploma level: the Royal College of Nursing, South Bank University and *Nursing Times*. The *Nursing Times* (a weekly journal) launched an open learning diploma-level programme in June 1993. This allows nurses to undertake modules with a CAT rating of 20 credits to obtain a diploma. Nurses without any level two CATs points can undertake the entire programme over two years to obtain a diploma. The *Nursing Times* diploma course consists of four modules:

Module	CATs Points at Level 2	Cost of Materials
Research	20	£110
Management	20	£115
Professional Development	20	£135
Teaching & Learning	60	£270

Nurses purchase the learning materials from *Nursing Times* and are linked to a study centre. The nurse pays for tuition at the study centre. An alternative method is for nurses to obtain telephone support and have their work marked by their appointed tutor.

South Bank University offer a range of options for completing the diploma.

The RCN Institute offer study materials, tutorial support and the opportunity to meet other students at their nation-wide study centres.

Further Information
Marian Phillips, Nursing Times, Open Learning, Porter's South, Crinan Street, London N1 9XW.

Distance Learning Centre, South Bank University, Borough Road, London SE1 1DW. Tel: 0171 928 8989.

Elisabeth Clarke, Distance Learning Co-ordinator, RCN Institute, 20 Cavendish Square, London W1M 0AB. Tel: 0171 409 3333.

Post-Registration Courses

There is a wide variety of post-basic courses for the registered nurse to choose from and full details are available from relevant national boards. Each national board produces a booklet giving details of all courses and training colleges.

One of the most exciting developments in nurse education has been the linkage of post-registration courses with diploma qualifications. This works in two ways: some new courses allow nurses to obtain a specialist qualification and a diploma or degree at the same time; others award CATs points as well as the relevant national board qualification. These CATs points can be used towards a Diploma or Degree in Nursing.

Degree-level Studies

Nurses who have 120 level two CATs can obtain a further 120 CATs points at level three to gain an honours degree in nursing. Many post-basic courses now carry credits at level three as well as level two. Level three CATs points gained can be used towards a degree.

You may prefer to enrol on a degree course. There are three basic options:

1. A year's full-time study
2. Two or three years' part-time study
3. Open learning options, normally two to three years.

Full-time and part-time courses are offered at most universities and nursing colleges. The RCN Institute and South Bank University offer distance learning options. *Nursing Times* is to launch a distance learning degree in 1997. The format will be similar to that of the diploma.

Costs vary from centre to centre but are in the Region of £2,000 to £3,000 for the course.

Master's Degree Studies

Most universities specify that you must have an honours degree before you can begin a Master's degree. A Master's degree is made up of 120 level M CATs points. Full-, part-time and distance learning options are available.

A full-time degree takes one year to complete. Part-time courses last two or three years. Distance learning can be undertaken over two to five years. There are a growing number of Master's degree courses. Many are advertised in the nursing journals and your local university can supply details of their MSc courses. The RCN Institute and South Bank University both offer open learning courses at Master's level. The cost of a part-time Master's degree is approximately £4,000.

Combining Theory with Practice: the ENB Higher Award

The National Board's Higher Award is a new post-registration professional and academic award introduced by the ENB in April 1992. It can be gained by completing a modular programme of professional development in which the nurse improves her skills and increases her expertise in ten key areas. Credit may be awarded for qualifications already gained. Nurses will be able to tailor a modular programme which takes account of their own interests, skills, practice and work area. The award can be studied through colleges of nursing and universities. Details of centres which have been approved to offer the higher award are available from the relevant national boards.

Nurses who achieve the higher award will have a deeper understanding of the professional and health issues which relate to nursing.

Working as a Nurse Teacher

In the past nurse teachers were experienced nurses who undertook a one-year full-time course and became registered nurse tutors. Changes in basic and post-basic nurse training now make the minimum qualifications for nurse teachers an honours degree. This is usually combined with a postgraduate diploma or Master's degree in Nurse Education.

Further Reading

Portfolio Development and Profiling for Nurses (1996). Roswyn Brown, CHS.
Profiles and Portfolios. A Guide for Nurses (1996). Cathy Hull and Liz Redfern, Macmillan.

Chapter 6
Returning to Nursing

Until recently, many nurses left nursing to bring up a family and failed to return because of the difficulties they experienced when they made the attempt. The profession used to replace itself every 7.5 years. Research undertaken in 1986 calculated that it cost £30,000 to train a registered nurse and some recent estimates have put the figure at double that because Project 2000 nurses contribute fewer rostered hours' service than traditionally trained nurses during their training.

This has now been recognised as an appalling waste of talent. It also deprives patients and less experienced staff of the expertise and skills of mature nurses. There has been a change in attitude within all areas of nursing and nurses returning after a break in practice are made more welcome than before.

Refresher Courses

Nurses currently returning to practice do not have to undertake a mandatory refresher course but this will change in the year 2000 when all nurses who have not practised for five years will have to take a UKCC-approved refresher course before they can resume practice. Many colleges already offer return to practice courses. You can obtain details from your local college or the relevant national board.

Supervised Practice

UKCC guidance states that newly qualified nurses, nurses working in areas where they have no previous experience and nurse returners should have a period of help and support from a more experienced nurse who has been working in the work area for some time. This nurse is called a mentor and the period of mentorship will normally be six months, though it may be longer for part-time staff.

Flexible Working Hours

Many nurses with children find it difficult to work the full range of shifts. The increase in day surgery and community care has resulted in more opportunities to work part time. If is often possible to start work at 9 or 9.15 am which allows parents to drop their children off at school before starting work. It is worth enquiring about such posts when you are considering a return to nursing.

Some nurses have found that certain posts, such as nurse teaching or practice nursing, allow them to combine motherhood with a career.

The majority of part-time posts are in the lowest grades of nursing. Some mothers are happy to mark time and have a fairly undemanding post while the children are young but others feel that they want a position of greater responsibility combined with flexible hours.

Job sharing can offer a solution. This is a method of working where two nurses share one job and work flexible hours between them. These job-share posts are often more senior than posts available on a part-time basis. The Royal College of Nursing produce a leaflet on job sharing which can be obtained from their regional offices or from headquarters. There is also an organisation called New Ways to Work which provides help and support and produces leaflets on job sharing and flexible working patterns.

Royal College of Nursing, 20 Cavendish Square, London W1M 0AB; Tel: 0171 409 3333

New Ways to Work, 309 Upper Street, London N1 2TY; Tel: 0171 226 4026.

Arranging Childcare When You Return to Work

A number of hospitals now run holiday play schemes which provide care, fun and play for school children while their mothers work. It is worth enquiring about this when you are applying for a job. If the hospital does not run a holiday play scheme, there may well be one in your area. You can obtain details by contacting Kids Club Network; they have local branches all over the UK.

Kids Club Network, 279-281 Whitechapel Road, London E1 1BY; Tel: 0171 247 3009.

Taking the Plunge

Many nurses lose confidence after they leave nursing. They feel that they have forgotten all their skills and that nursing has changed beyond recognition. Although there have been many changes in nursing, human nature has not changed. Patients still require skilled nursing care and nurses with experience of bringing up a family can provide top quality care. They understand how parents feel when their child is ill, how devastated adults can be when their elderly parents are ill; they have not only theoretical professional knowledge of childcare but also practical experience as a parent. Yes some things have changed: nurses use plastic forceps instead of metal, rubber latex gloves are used more often and solutions for cleaning wounds now come in sachets instead of in bottles, but the essence of nursing and caring remains unchanged.

It is difficult and frightening at first to return to nursing. You must reorganise your life so that the household chores you had all week to do are fitted into a few hours, but you will once again have the benefits of a fulfilling career. The only thing to fear in returning to nursing is fear itself. You will be nervous and uncertain at first but with the support of colleagues, many of whom have been in the same position, you will regain your confidence.

Case Study
Jacqueline Roy returned to nursing three years ago. She was a ward sister before leaving to bring up three children and describes her feelings on returning to nursing.

> Although I had nursed a lady in her own home for some years, I had not worked with other nurses in a formal setting for more than ten years. I was very nervous and felt that I had forgotten everything I had ever learnt. I enrolled on a return-to-nursing course but it concentrated on theory and my greatest fears were about my practical skills. The course did not provide any supervised practice.
>
> I returned to nursing after seeing a part-time job advertised in a local paper which stated that the job was suitable for a nurse returner. I needed a great deal of support in the first weeks, not because I had forgotten my skills, but simply because I had lost all my confidence. If I had not had the support of colleagues, I would have been unable to make a successful return to nursing. I'm now glad to be back and looking forward to doing a specialist course in the future.

Jacqui has been promoted three times in the last three years and now manages a nursing home.

There is an organisation called the Working Mothers Association which provides help, support and advice to women from all walks of life who are returning to work. The WMA has many local branches and local contacts can give advice on the availability of childcare as well as providing support. They publish a book called *The Working Parent's Handbook* and can be contacted at:

Working Mothers Association, 77 Holloway Road, London N1 8JZ; Tel: 0171 700 5771.

Single Parents

Single parents can find it especially difficult to return to work if they do not have family nearby who can provide help with childcare. The single parent with children under school age can find it difficult to afford to pay for childcare from her salary. The organisation Gingerbread has many local branches. Some of them run nurseries which charge much less than local childminders and nurseries. Gingerbread also offer advice and support for single parents. They can be contacted at:

Gingerbread, 33 Wellington Street, London WC2E 7NB.

Further Reading

How Not to be the Perfect Mother (1986). Libby Purves, Fontana, London.
The Working Parents' Handbook (1994). Working Parents Association, London.

Further Information

The National Childminding Association, 8 Mason's Hill, Bromley, Kent BR2 9EX. Tel: 0181 464 6164. Provide information and leaflets about childminding.

National Childbirth Trust, Alexandra House, Oldham Terrace, London W3 6NH. Tel: 0181 992 8637. Run nanny share registers and support groups for working mothers through local branches.

Chapter 7
Working in the Community

As patient care has moved from hospital to the community the opportunities for nurses have grown: to work as a district nurse, a health visitor, a school or family planning nurse or as nurse specialist such as a continence adviser or wound care specialist, or a community midwife. There are also opportunities to work as a practice nurse. The implementation of the Community Care Act 1993 has led to community nurses working with acutely ill and convalescent patients as well as their traditional clients.

Community Nursing

District nursing sisters have responsibility for planning and supervising care. They lead a team of staff: nursing auxiliaries, trained district enrolled nurses and registered general nurses who have not had a district nurse training. They visit and assess patients who require care in their own homes. They work with each patient to develop a plan of care and delegate aspects of that care to different members of their team. They aim to enable the person to become as independent as possible.

In the past district nurses worked mainly with elderly people but this is changing. Hospitals are now discharging patients more rapidly than before. Developments in surgical techniques have led to many patients having surgery as 'day cases'. The district nurse visits day case patients in their homes and checks on their progress. She may run special clinics where patients can come to have stitches removed or wounds dressed.

Many NHS Trusts have set up special intensive care schemes to care for patients at home, usually called 'hospital at home'. Patients are discharged early from hospital and district nursing staff visit the patient's home several times a day for a period to provide care. Hospital at home schemes normally operate for about two weeks as this is the normal time it takes a patient to recover.

Community Nurse Education
All district nurses are registered general nurses, who are usually expected to have at least a year's experience before they can begin education to prepare them for roles in community settings. There are now three routes to community nurse qualifications:

1. A degree awarded by a UK university and approved by the relevant national board as a specialist qualification recordable with the UKCC.
2. A Diploma in Higher Education awarded jointly by the university and the relevant national board and recordable (with health visiting registrable) with the UKCC.
3. A modular Certificate and Diploma in Professional Practice – both are recordable with the UKCC.

Degree and diploma-level courses last one year full time. Increasingly, students can study for these qualifications on a part-time basis. Part-time degree and diploma level courses can be completed in two to five years. Students must have a diploma before they can study at degree level.

Courses are available in:

- District Nursing
- Health Visiting
- Community Psychiatric Nursing
- Learning Disability Nursing (Community Practice)
- Occupational Health Nursing
- School Nursing
- Practice Nursing

Nurses with community nurse qualifications are paid at a minimum of G grade.

District Nurse Training for Enrolled Nurses

Enrolled nurses with a district nursing qualification (DEN) can complete a dual conversion course. These courses enable the EN to convert from EN to RN and DEN to DN. They are full time and last one year.

Master's Level Courses

A number of Master's level courses are now being introduced. They are for qualified community nurses who wish to gain additional knowledge and expertise.

Working as a District Staff Nurse

Qualified RGNs who have not undertaken the district nursing course often work as community staff nurses. Many nurses choose to work as a staff nurse so that they can gain insight into the role of the community nurse. Many go on to study for a community nursing qualification. The rates of pay vary from one area to another; some nurses are paid on D grade; others on E or F.

Specialist Community Nurses

In recent years the number of specialist community nurses has increased. Many are district nurses with specialist skills in particular areas, for example in wound care. These specialist nurses help and advise community nurses who are caring for patients with particular needs.

Community nurses specialising in wound care have achieved impressive results, treating patients with leg ulcers in leg ulcer clinics. Continence advisers have successfully helped many people to regain continence. Specialist district nurses are often paid on an H or I grade.

Community Psychiatric Nursing

Community psychiatric nurses are registered mental nurses (RMN) with specialist training in community work who care for and support patients with psychiatric problems in the community. One national board course, ENB 812, remains. The 36-week long course aims to enable the RMN to provide appropriate community care. It is offered as a series of study blocks and the nurse works in the community as a CPN while doing the course. Increasingly, post registration is moving to diploma and degree level and this course may be phased out.

Health Visitors

Health visitors are RGNs with specialist training. They work in the area of health promotion and do not provide 'hands on' nursing. While many health visitors and their employers have concentrated on elderly people and children under five, the health visitor is in fact trained to promote health for all. The changes in community care may mean that the health visitor's role is expanded to cover all ages. Many specialist health visitors promote healthy life styles and work with elderly people in the community. Falling birth-rates and increasing numbers of older people may mean that work with elderly people becomes a central part of the health visitors role over the next ten years.

Health Visitor Education
Education for health visitors is now at diploma or degree level. They are trained at colleges of further education or at universities. On qualification, health visitors are paid on a G grade. More experienced and senior health visitors are paid on higher grades.

Career Opportunities for Health Visitors
It is normal for newly qualified health visitors to gain two years' experience before beginning any further specialist training. Then the experienced health visitor can go on to train as a field work trainer, teaching, training and supporting other health visitors and health visitor students in the community. The health visitor may want to train to become a lecturer in a college or university or she may prefer to become a health visitor manager and lead her own team.

School Nursing

School nurses are employed to monitor the development of children and to educate them about health issues. Each one is usually responsible for several schools. The number will vary depending on how many pupils each school has.

The school nurse checks children's height and weight, carries out hearing tests and alerts the school medical officer to any problems which she detects. She also has an important role in advising and supporting parents whose child has a health problem. She works with pupils from the ages of 3 to 18, advising children and their parents (especially in the case of younger children) about health issues.

She works closely with professionals such as speech therapists, physiotherapists and others who may be treating a child attending one of the schools she is responsible for.

Training as a School Nurse
Only nurses trained in adult or children's nursing can become school nurses. Registered nurses are required to have a few years' experience before applying for school nurse training, which is now at diploma and degree level.

Career Opportunities
There are opportunities for school nurses to enter management or to become nurse teachers after specialist teacher training.

Practice Nursing

Practice nurses do not work directly for the NHS but are employed by general practitioners to provide nursing services to their patients. There are over 20,000 practice nurses in the UK, mostly working part-time, and this is a new and rapidly growing speciality. The role of the practice nurse is as broad as she and her employer wish to make it.

The work is divided into two areas. The first is working as a clinical nurse and undertaking duties such as taking blood samples, removing stitches, doing dressings and syringing ears. The second, more radical role is in health promotion.

The practice nurse has in many ways begun to fill the gap in health promotion which arose because health visitors tended to concentrate on working with the under-fives. The practice nurse may run a series of clinics which would help overweight people to lose weight, smokers to give up smoking, etc. She may run clinics which would help asthma sufferers to cope with their condition. She may visit elderly people in their own homes and undertake over-75 health checks. She will inform the GP if she detects any problem which requires medical intervention or the services of another professional such as a physiotherapist.

Working as a Practice Nurse
Enrolled nurses can work as practice nurses but the UKCC states that they may 'only undertake a limited range of duties'. The Royal College of Nursing described this statement as 'somewhat restrictive' and have issued guidance which suggests that enrolled nurses can fulfil a fuller role as a practice nurse provided they can demonstrate their competence.

Registered general nurses often work as practice nurses but some health visitors and district nurses have found work as practice nurses attractive. Practice nursing has proved a popular option for nurses with domestic commitments because the working day usually starts around 9 am and there are flexible finishing times. Most practice nurses work one or two evening sessions a week and usually alternate Saturday mornings.

Practice Nurse Education
It is still possible to work as a practice nurse without undertaking any specialist education. However, nurses have a professional duty under the UKCC code of conduct not to perform a role they are not competent to carry out. Some colleges of nursing offer a ten-day practice nurse course, but increasingly practice nurse education is at diploma and degree level. Practice nurses can attend study days which deal with specific areas of health promotion or diseases to fulfil their PREP requirements after qualification.

Pay and Conditions
Practice nurses are employed by general practitioners and their pay and conditions are a matter of negotiation between the nurse and her employer. Salary scales vary from F to H grade according to the role the practice nurse undertakes. There are no standard terms for such areas as holiday pay, maternity leave or sickness benefit. The Royal College of Nursing produce a leaflet giving details of recommended salary scales and terms for practice nurses. They also provide a large detailed booklet *Guidance on the Employment of Nurses in General Practice*, available free to nurses who are members of the RCN.

Chapter 8
Working in the Independent Sector

There has been considerable growth in the number of people cared for in the independent sector in the last ten years. More elderly people are being cared for in residential and nursing homes as the number of elderly people being cared for in NHS continuing care units has fallen. There has also been modest growth in the acute independent sector; that is in the private hospitals offering mainly surgical care.

Working in Private Hospitals

Private hospitals charge for the care which they give patients; most patients have private health insurance which covers the cost of their treatment. Private hospitals tend to concentrate on offering surgical treatment or investigating acute medical problems. They usually demand high levels of experience and expertise from their staff and the atmosphere of such units is different from that of the NHS.

Pay and Conditions in Private Hospitals

In the past, pay and conditions in these units were far superior to those offered to NHS nurses but this is no longer the case. Nurses in these units are now usually paid on the same salary scales as NHS nurses.

Private hospitals tend to have higher staffing levels than their NHS equivalents and are usually much smaller, typically having only about 50 beds. Many nurses prefer to work in these units because they feel that the higher staffing levels give them an opportunity to deliver top quality care to their patients. Some nurses enjoy working in a small unit instead of the increasingly large NHS units.

Elderly Care in the Independent Sector

Residential homes are not legally obliged to employ qualified nurses; however, many residential homes are run by a qualified nurse.

Nursing homes offer skilled nursing care on a 24-hour basis for over 222,000 elderly people in the UK. There are many opportunities for skilled nurses in this area. The work can be rewarding and flexible hours are often available which allow nurses to combine the job with domestic responsibilities. Residents may include elderly people normally cared for at home while their families have a break. This is known as respite care.

It also involves rehabilitating elderly people and enabling them to return home after a major illness as well as providing continuing care for individuals who live permanently in the nursing home.

Many employers prefer nurses with specialist qualifications in elderly care such as ENB 941 or 978. Specialist qualifications in areas such as continence promotion, palliative care and rehabilitation nursing equip nurses with important skills to enable them to provide quality care for elderly people.

Pay and Conditions in Nursing Homes

In the past salary scales have been lower than in the NHS. However, the current shortage of registered nurses and increased competition between employers are raising salary rates. Annual holidays, sickness pay, maternity leave provision, etc are fixed in larger nursing homes which are members of groups. In smaller homes these matters are often a subject of negotiation. Recommended salary scales are set out in Table 7.

Table 7. Recommended Salary Scales for Nursing Home Nurses

Grade	Hourly Min-Max	Annual Rate	Notes
C	£5.54-£6.56	£10,832-£12,834	Enrolled Nurse grade
D	£6.35-£7.27	£12,418-£14,206	Junior Staff Nurse grade
E	£7.27-£8.42	£14,206-£16,454	Staff Nurse grade
F	£8.06-£9.87	£15,737-£19,306	Senior Staff Nurse/Junior Sister
G	£9.10-£10.53	£18,582-£21,500	Sister
H	£10.62-£12.15	£20,766-£23,750	Senior Sister/Manager
I	£11.76-£13.22	£22,922-£26,053	Manager/Senior Manager

Salary enhancements

The RCN recommends that duties between 8 pm and 6 am weekdays and on Saturdays midnight to midnight should be paid at time plus 30 per cent. Sundays should be paid at time plus 60 per cent. In practice enhanced rates are rarely offered but hourly rates may be higher to take account of this. London allowances (details in Chapter 3) should be paid to staff working in the London area.

Continuing Your Professional Development

At present no specialist post-basic qualifications are required for nurses working in nursing homes. However, the RCN have just produced a booklet, *Nursing Homes – Nursing Values* (free to members), that recommends nursing home managers should have specialist qualification in elderly care such as the ENB 941. South Bank University plan to launch a specialist course in nursing home nursing in 1977 which will combine ENB qualifications with a Diploma or Degree in Nursing Home Nursing.

Further information is available from:

Allan Hicks, Course Director, Redwood College of Health Studies, Harold Wood

Education Centre, Harold Wood Hospital, Gubbins Lane, Harold Wood, Romford RM3 0BE. Tel: 0171 928 8989.

Nurses who work in nursing homes often feel isolated from developments and changes within their field and are keen to share information and experiences with other elderly care nurses. Nurses can read professional journals in order to keep up to date. Several specialist journals are designed especially for nurses working with specific groups of patients, and the one for nurses working with elderly people is called *Elderly Care* and can be obtained on subscription from the publishers of *Nursing Standard*.

The Royal College of Nursing has two special groups for nurses working in elderly care; they are called ACE and FOCUS. ACE is concerned with general elderly care while FOCUS is concerned with the mental health of elderly people. Nurses can join these groups free of charge if they are members of the RCN and receive newsletters several times a year, as well as details of study days and courses which relate to the care of elderly people.

Careers in Nursing Homes

There are opportunities to work as a staff nurse, sister, manager or even as a manager of a group of homes. The larger nursing home groups employ nurse lecturers to provide in-house continuing education and in many such groups there are opportunities for enrolled nurses to undertake conversion courses.

Career opportunities within this expanding sector are growing rapidly as the nature of nursing home work changes. Nurses working in nursing homes now require a greater range and level of skills than ever before.

Chapter 9
Nursing Abroad

A recent survey revealed that 10 per cent of UK nurses had worked abroad at some time and that a further 20 per cent were keen to do so. Many newly qualified nurses want to work abroad but most employers require two years' post-registration experience. Some countries prefer UK nurses with specialist qualifications while others do not recognise these.

There are many reasons why nurses wish to work abroad; some use work as a way of financing a working holiday; some wish to gain a period of experience in another country; some wish to move to another country permanently and others to undertake work on a voluntary basis in a developing country.

The most popular destinations are the United States of America, Canada, Saudi Arabia, European countries and developing nations.

Nurses considering working abroad must think carefully about their decision. There are many personal and professional rewards but the path to working abroad is more difficult than before and gaining registration in another country can be expensive. It is important to learn as much as possible before applying for posts abroad.

The International Department of the Royal College of Nursing produce a number of booklets giving further information.

Cultural Differences

The term 'culture shock' describes the reaction of individuals when they encounter a different culture and working environment in another country. It is often assumed that this relates only to the differences between developing countries and industrialised societies but every move to another country and a different working environment involves a certain amount of culture shock. The degree of this shock varies from country to country and can be minimised by learning as much as possible about the language, customs and culture of the country you plan to work in before you go. Worth reading is *Working Abroad: The Daily Telegraph Guide*, published annually by Kogan Page.

Career Planning

Many nurses wonder if there is a 'right time' in their working lives to work abroad and what effect this will have on their careers if they later return to the UK. The choice of a right time must be made by the individual concerned but many nurses who work abroad for a few years tend to do it either at the beginning or the end of their careers.

A nurse may work abroad in the early years of her career before she has begun to climb the career ladder. She usually gains a wealth of experience which enriches her practice on return. Once she has begun to ascend the career ladder, a period of work abroad can be risky, as in the current UK job market, she may be unable to obtain a post at the same level as the one she left. A nurse in a senior position can have much to lose from a period of work abroad. Some nurses work abroad because their partner's job has led to such a move; they decide that their career will take second place to their partner's for a time.

Nurses at the end of their career have often a wealth of experience at senior level in either management or teaching and this experience is valued in many countries. These nurses have little to lose in leaving the UK to work abroad and can gain a great deal. Many senior nurses choose to end their careers with a few years abroad in senior posts, especially in the Middle East.

Length of Stay

Nurses intending to work abroad for the first time have little idea of what it will really be like. In the first instance, it is advisable to sign a contract for a year which can usually be renewed. It is possible to obtain locum posts for short-term contracts; these are usually for three or six months and they allow the nurse to decide if she would enjoy working in the country for a longer period. Locum posts are advertised in the nursing press and are available in Europe and the Middle East. The contract usually includes return air fare and living accommodation.

Working in the US

The nursing journals used to be crammed with advertisements a few years ago, and UK nurses could pick and choose between employers in the US competing to recruit them. The recession there has resulted in a lower nurse turnover and the US federal government have introduced strict rules about overseas recruitment so there are now fewer opportunities for UK nurses.

There are no national rates of pay and conditions of service such as the UK Whitley system. There are wide differences in the rates of pay and allowances, so it is advisable to enquire about the terms in different hospitals in order to get an idea about local rates. It is not unusual for overseas nurses to be allocated to night duty for the first year of work in a US hospital so it is wise to enquire about shift patterns. Nurses in the US may receive between two to six weeks paid holiday per year but the average is three to four weeks. Nurses should enquire about holiday

Nursing Abroad

entitlement and also if paid flights home to UK are part of the contract. Some hospitals expect the nurse to work for a full year before she is entitled to any holiday, so it is important to check this. It can be difficult working for a whole year in a strange country without a break.

In the US, there is only one type of registered nurse (RN) and no equivalent qualifications to RSCN, RNMH or RMN; only nurses who are RGN or RN (adult branch) are eligible for US registration. There is no national board for nursing; each state has its own board and its own rules for employing overseas nurses.

The nurse wishing to work in the US must have a valid work permit, called an HA-1A visa. It can only be applied for by the nurse's US employer.

CGFNS Tests
The nurse may have to sit a test known as the CGFNS (Commission on Graduates of Foreign Nursing Schools); some states such as Florida and California do not require this. Nurses are usually informed of CGFNS requirements by their prospective employers. It is possible to sit the CGFNS examination in the UK. Nurses who are successful are given an interim licence to practise in the state and are booked to sit another examination called the NCLEX-RN. No examinations are required for UK trained nurses wishing to practise in the state of Arkansas.

Permanent Migration
Nurses who have registered in the US and obtained a work visa can apply for citizenship and these applications are generally welcomed. It is important to note that nurses who do not possess at least a basic degree in nursing will find it extremely difficult to obtain promotion even to head nurse (sister) level.

Further Information
The Royal College of Nursing's International Department produces information leaflets giving details of US state boards and professional associations as well as useful advice. These are available free to members.

Working in Canada

There is currently a recession in Canada and as some Canadian nurses are unemployed, all overseas recruitment has been halted. Employers can no longer advertise for overseas nurses, recruit them or obtain work permits for them to work in Canada. Only nurses who intend to emigrate to Canada permanently will be able to find work in the present economic climate.

Nurses who wish to work in Canada must have a work visa. Further information is available from the Canadian High Commission, Immigration Division, 38 Grosvenor Street, London W1X 0AA.

Registration in Canada
Nurses who wish to work in Canada must be registered there, in the

province they wish to work in. They must first satisfy that Canadian province that their experience includes, medical, surgical, obstetric, paediatric and psychiatric nursing. Nurses with satisfactory experience are allowed to sit the examination administered by the Canadian Nurses Association Test Service (CNATS). This examination can only be taken in Canada and is run four times a year at most. Individual provinces can give details of when these examinations are run.

Employment opportunities for RMNs are more limited as only four provinces, Alberta, British Columbia, Manitoba and Saskatchewan, maintain separate registers for psychiatric nurses. The examinations for psychiatric nurses are run twice a year. It may be possible to obtain a temporary licence to practise in the meantime. Nurses should contact individual provinces to check this.

CNATS Examination

This is an examination of nursing knowledge and aims to test theory and practice. The examination takes place over four days. It is a multiple-choice examination and nurses must complete four sessions where there are 120-130 questions in each section. The questions cover all aspects of practice: paediatrics, obstetrics, psychiatric nursing, anatomy, physiology, pharmacology, nutrition, growth and development.

Temporary Licences to Practise

The rules governing the issue of a temporary licence to practise vary from province to province. Some provinces grant a temporary licence while the nurse is waiting to sit the CNATS examination; others only grant it after the nurse has taken the examination and is awaiting the results. Nurses can obtain more information by checking with the individual province.

Working in Australia

Australia is the most popular destination for nurses wishing to work abroad. Australia recruited large numbers of UK nurses until a few years ago but the situation has now changed. The recession has led to cutbacks in the health service there. Many qualified nurses in Australia are now unemployed and there are no guarantees of a period of employment to consolidate practice for newly qualified nurses. Nurses can write to hospitals asking about vacancies but they are usually asked to phone the hospital when they arrive in Australia. Nurses can find employment through nursing agencies provided they have work visas and are registered to work in the state they intend to work in. All newly registered nurses in Australia are qualified to degree level. Nurses considering permanent migration will have limited career prospects if they are not so qualified.

Work Visas

There are three types of work visa. The first is a temporary work visa and nurses who are offered a post are helped to obtain this. Nurses who wish to emigrate to Australia require a migrant visa. It can take up to a year to obtain one and applicants are charged fees for processing, medical examination and registration. Working holiday visas are usually intended for people aged 18-25 but under certain circumstances can be issued to people who are up to 30 years old. Working holiday visas are only issued once; they last for up to a year and cannot be renewed. Applicants must be able to provide evidence that they have sufficient funds to pay for their air ticket home after the holiday. The prospects of finding sufficient temporary work as a qualified nurse are so poor that nurses are currently advised that they should be prepared to do work other than nursing to finance their stay.

Registration in Australia

Nurses applying from the UK must apply for registration to the Australian Nursing Assessment Council (ANAC). It can take months for nurses to register when they arrive in Australia and nurses *may not work as qualified nurses until they have effective registration*. ANAC forms can be obtained from the nearest Australian consul. ANAC charges a processing fee of A$100 and £20 to verify UK registration.

Enrolled nurses who have completed the conversion course will experience great difficulty in obtaining registration if they do not have a minimum of one year's post-conversion experience and should take advice before sending off money for registration as this is not normally refundable.

Each Australian territory or state currently has its own registration board but there are plans to introduce an Australian Nursing Council which would be responsible for all nursing registrations.

Recognition of Post-Basic Qualifications

At present some states and territories recognise post-basic certificates while others do not. The introduction of the Australian Nursing Council should clarify this situation.

Further Information

Australian High Commission, Australia House, Strand, London WC2B 4LA

Australian Consulate, 2nd Floor, Hobart House, 80 Hanover Street, Edinburgh EH2 2DL

Australian Consulate, Chatsworth House, Lever Street, Manchester M1 2DL.

Working in Saudi Arabia

There are fewer posts available in Saudi Arabia than before but there are more opportunities to work in Saudi Arabia than in other countries at present.

Saudi Arabia is a Muslim country, and its laws are based on the Koran. The nurse who works in Saudi Arabia is entering a completely different culture with a different set of values and norms from those which she has been brought up with. Alcohol is banned; it is not possible to buy pork; women do not drive. A woman must conform to a strict dress code which dictates that she wear a knee-length (or preferably ankle-length) skirt, cover her arms at least to the elbow and dress modestly when in public. Many Saudi women are veiled and only reveal their faces to members of their own sex and their own immediate family.

Most overseas staff live in compounds where flats and swimming, sports, recreational and shopping facilities are provided. Life is usually much more relaxed within the compound. The nurse must carry identification with her at all times.

She will usually be unable to speak Arabic and her patients will be unable to speak English. Every ward and department has at least one interpreter and there are on-call interpreters; some of the staff may have come from other Arab countries and will be able to help. Most hospitals offer Arabic classes for nurses keen to learn the language.

Working Conditions
Saudi hospitals are usually modern, well equipped and well staffed. Salaries are normally higher than in the UK and are tax free to people who live in Saudi Arabia for one year or more.

Accommodation is usually supplied free of charge and there are seldom charges for electricity or water. Most flats come equipped with a telephone and all calls made within the country are free. International calls are very expensive.

The working week is usually longer than in UK and a 45-hour working week is common. Holidays vary from company to company; some offer one month every six months while others offer less. Flights home are usually included every six months. There are opportunities to travel within the country or to fly to neighbouring countries such as Egypt or Bahrain for a long weekend.

Work Permits
These are obtained by the employer for the nurse and it is not normally possible for nurses to change their jobs in Saudi Arabia. They must return to the point of hire at the end of their contract and apply for another post.

Further Information
The Kingdom by Robert Lacey is an interesting personal account of life in Saudi Arabia (Hutchinson, 1981). The RCN International Department can provide further information on working in Saudi Arabia and other Middle Eastern countries.

Working in Europe

The introduction of freedom of movement within the European Community, and EC directives which have allowed mutual recognition of nursing qualifications throughout the EC, have led many nurses to consider working in Europe.

The only obstacle to employment in Europe is fluency in the language of the host country. A command of the language to at least A level standard is required.

Posts are advertised in the major nursing journals. It is important not to accept the first post offered but to make enquiries to several different agencies or employers in the area. This will give you an idea of the work conditions and rates of pay.

Further Information
It is possible to seek work through:

The Employment Service, Overseas Placing Unit (OPS3), c/o Moorfoot, Sheffield S1 4PQ.

The OPU also offers advice and guidance on working in Australia, Norway and Sweden.

Voluntary Work Abroad
Many newly qualified nurses are keen to work abroad in a voluntary capacity in developing countries but developing countries usually need the skills of more experienced nurses. A minimum of two years' post-registration experience is usually required and extra qualifications such as midwifery, health visiting or teaching are valued. It is difficult for enrolled nurses to obtain voluntary posts abroad.

Life in Developing Countries
The nurse working in a developing country is paid an allowance to cover daily living expenses and is provided with free accommodation. She often works in partnership with a doctor or other professionals but may work on her own for long periods of time. She will live among the local community in simple accommodation with people who have a different culture, language and values. She will have to make great adjustments and be willing to learn as well as to teach.

Long-term Commitment
Long-term contracts are usually for one or two years and involve the nurse giving up her job and often her home in the UK.

She is normally very well supported by the voluntary organisation she is working for and will receive in-depth training prior to departure. This training covers all aspects of politics and culture and usually includes lessons in the language of the host country. She is also taught how to adapt her UK skills to very different circumstances. There are now courses available in tropical nursing and primary care in developing countries.

She may return two years later to find her friends have gained promotion and material possessions while she has been working as a volunteer in a developing country. She will usually have gained a wealth of experience on return but may find that this is not valued by some NHS managers. The Royal College of Nursing have produced a booklet, *The*

Value of Overseas Nursing, which is free to members. It emphasises the value of overseas experience.

Further Information
Information on courses is available from:

Hospital for Tropical Diseases, 4 St Pancras Way, London NW1 0PE

Liverpool School of Tropical Medicine, Pembroke Place, Liverpool L3 5QA

Institute of Child Health, Tropical Child Health Unit, 30 Guilford Street, London WC1N 1EH

Oxfam Regional Office, 4 Replingham Road, Southfields, London SW18.

Short-term Commitment
This is usually a posting of three or six months to help in the case of a disaster or crisis. It is usually open only to those who have had a long-term posting: the part of the world the volunteer was posted to is not relevant but the skills which she developed during her post are vitally important. Nurses are usually contacted at short notice but NHS employees and employees in large organisations are usually paid their normal salary in addition to their jobs being kept open for them on return.

Further Information
The International Health Exchange, 8-10 Dryden Street, London WC2E 9NA

International Health Exchange produce the *IHE Magazine* which lists all health vacancies in developing countries. It has also developed a central disaster relief register. It keeps information on all relevant courses in the UK and runs its own short courses. For further information, send a large sae to IHE.

Ahrtag Ltd, 1 London Bridge Road, London SE1 9SG

Ahrtag produce a booklet called *Primary health care in developing countries: a guide to resources and information in the UK*. This costs £3.50 (October 1996) plus postage and lists 137 organisations in the UK which provide primary health care in developing countries; it also gives details of funding bodies, lists of relevant books, booklets and other information.

Chapter 10
Nursery Nursing

Nursery nurses care for children from birth to the age of seven years (eight years in Scotland); they normally care for healthy children while RSCN nurses care for sick children. Nursery nurses can work in hospitals assisting RSCNs to care for recovering children; in schools assisting teachers to care for young children in reception classes; in day nurseries; or in a child's home. They may work abroad. Nursery nursing, like any other career, can be as exciting, interesting and full of prospects as you wish to make it.

Training

Entry Requirements
There are over 190 colleges (some private) offering courses which lead to the Diploma in Nursery Nursing, awarded by the National Nursery Examination Board (NNEB). In theory, there are no minimum educational qualifications demanded for entrance to these courses but, because they are so popular, the colleges are demanding at least three GCSEs at grades A–C or S grades 1–3. Some colleges are asking for A levels or H levels in Scotland. All colleges state that they will consider candidates without these qualifications but prefer those who have them. Training can commence at the age of 16.

Types of Training
All NNEB students spend around 40 per cent of their time on theoretical work and 60 per cent gaining practical experience in a supervised or supported setting. Training is available from either colleges of further education or private colleges; both colleges offer exactly the same type of training and follow the same syllabus and both lead to the qualification of NNEB. Training in Scotland is slightly different (see below).

Colleges of further education offer a two-year full-time training or a three-year part-time training. The three-year part-time diploma is often studied by individuals who have jobs as unqualified nannies or nursery workers and wish to gain a professional qualification. Students on the part-time courses are often slightly older and have practical experience of working with children.

Private colleges offer NNEB training on a residential basis and the courses vary between 18 months and two years. Students must pay not only for their tuition but also for their board and lodgings while at these colleges.

Training in Scotland
In Scotland, diplomas are issued by the Scottish Nursery Nurses Board (SNNB) and the training is through a series of modules which cover periods of theoretical and practical work. Registration is issued when the student has completed the required number of modules and a placement of 140 days in two different types of nursery. These modules are called SCOTVEC National Certificate modules. They are available through 22 local authority colleges.

Working as a Nursery Nurse

The nursery nurse works in a series of placements during her training and usually has some idea of which area she would like to work in after qualifying. She has a number of areas to choose from and these are outlined below.

Working in a Hospital Nursery
A number of hospitals now have nurseries where the children of staff are cared for. These children are normally aged between three months and five years. Nursery hours vary and usually take some account of shift patterns worked by staff. They may open as early as 7 am and close as late as 9.30 pm. Children are not normally left in the nursery for more than nine hours as staff drop off their children when they arrive and collect them when they have finished their shift. Nurseries are not normally open at weekends but are often open during bank and Christmas holidays (though not on Christmas day or Boxing day). The working week varies between 35 and 37.5 hours and there is an element of shift working which will vary from nursery to nursery.

Working in a Local Authority Nursery
Local authority nurseries care for children from the age of six weeks until they begin school. In the past, many of the children cared for in local authority nurseries were the children of working or single parents and the work was similar to that in private nurseries. This has changed and it is now very difficult for parents to find their child a place in a local authority nursery: only children with 'special needs' are accepted. These children are usually suffering from mild to marked degrees of physical or mental handicap or come from extremely disadvantaged backgrounds. Some may have suffered physical abuse or come from families where one or both parents have severe physical or mental problems or a problem such as alcohol abuse. The number of children with severe behavioural problems in local authority day care has risen rapidly.

The job will involve working with other professionals such as doctors, psychologists and psychiatrists. Some nursery nurses find this work challenging and rewarding while others find themselves unprepared for this role without further specialist training.

Local authority nurseries are normally open from 8 am until 6 pm on a Monday to Friday so there is some element of shift work involved. Nursery nurses normally have five weeks' holiday, do not work bank holidays and have a week's holiday over the Christmas period.

Working in a Private Nursery

Nursery nurses normally care for children from the age of 18 months to five years in private nurseries. The majority of children attend private nurseries because their parents are at work during the day. Many of the children are supported by local authorities who often provide working single parents with a childcare allowance to allow them to place their children in private nurseries. This enables local authorities to provide care for children with 'special needs' in their own nurseries.

Nursery nurses tend to care for a small 'family group' of children and often supervise NNEB students who are placed at the nursery to gain practical experience. Working hours are usually Monday to Friday, and nurseries open from 8 am until 6 pm. Nursery nurses normally work a 40-hour week so there is some shift work involved. Between four and six weeks' holiday per year is usual and nurses do not normally work on bank holidays.

Working in a Local Education Authority Nursery

Local education authorities often provide nursery education for children from the age of three onwards. These classes are commonly for two or three hours per day and there are two sets of pupils: the morning children and the afternoon children. This is often the first time these children have been away from their mothers on a regular basis and the nursery nurse works with qualified teachers to help these young children to learn through play. The work is carried on in an informal atmosphere and there are lots of activities such as finger painting and water play. There is also a fair amount of clearing up to do between sessions.

Nursery nurses usually work a 36-hour week and have all the school holidays off.

Working in a School

Many local education authorities now admit children to school at 'rising five' – that is some time after their fourth birthday – and have special reception classes for these very young children. These reception classes commonly employ a qualified teacher and a nursery nurse.

The nursery nurse is uniquely qualified to help very young children to make the transition from nursery to school. She helps them learn to cope during this very important year and her knowledge of child development is of great help to the teacher who might not otherwise have time to note early signs of problems.

Nursery nurses in schools normally work a 35-hour week and have all school holidays off.

Working in Hospitals

Nursery nurses often work in three different areas within hospitals; the maternity unit, the special care baby unit and the children's wards. In some hospitals, they also work in the accident and emergency unit.

In maternity units nursery nurses work with midwives and mothers. They assist midwives to care for newborn babies when mothers are too unwell to do so. They assist and advise mothers in caring for newborn babies. This is especially important with first-time mothers who may lack confidence; sometimes even experienced mothers can feel a little nervous when handling a newborn and nursery nurses can help mothers to gain confidence.

Nursery nurses often assist sick children's nurses and midwives in caring for sick and premature babies in special care baby units. They also provide advice and support to parents during this distressing time.

The nurses also have an important part to play in caring for sick children in hospital. Unfortunately, some children suffer from diseases which necessitate prolonged hospital treatment or frequent hospital admissions. Nursery nurses can provide care for children who are on the mend and can help the child to adjust to this period in hospital by playing and organising activities which the child can enjoy although ill, weak or even confined to bed.

Nursery nurses in hospitals work a full range of day shifts and normally work seven days a week including bank holidays. The normal working week is 37.5 hours with five weeks' annual leave and a further ten days' leave in lieu of bank holidays.

Working as a Nanny

A nanny normally cares for one or more children from a family while the child's parents are at work. She may live in the family home or nearby. The working hours are usually longer than in other posts but she may have time to herself during the day if the child she cares for is at school, nursery or playgroup.

The rates of pay vary according to the part of the country; nannies working in London tend to earn more. Daily nannies earn more than those who live in (though live-in nannies have free accommodation and food) and some employers pay more than others. Daily nannies earn from £80 to £250 per week and live-in nannies from £60 to £220 per week. Holidays are usually a matter of negotiation.

Working Abroad

There are opportunities for nannies to work abroad. These are often with families whose work involves either overseas travel or who live abroad. These posts are popular and pay can be as much as £350 per week and accommodation and board are provided by employers.

There are also opportunities for nursery nurses to work abroad with package tour operators. In many popular resorts, package tour operators provide children's clubs. These clubs provide play, excursions, treasure hunts and other activities for children from the age of two or three for a few hours each day while parents have a break. There are normally two sessions of three hours per day and children attend either morning or afternoon sessions.

The nursery nurse employed by a tour operator normally works in two resorts over the year; she may work in a Mediterranean resort from April until October and in a winter sun resort from October until April. Weekends are usually free and the normal working week is around 40 hours. The pay is between £7,450 and £9,450 per year, all found, and there are usually two or three paid return flights per year and six weeks' holiday. There are also generous concessions on overseas flights and holidays after a qualifying period of one year.

Nannies can also work abroad for short periods on such schemes as 'Camp America'. In the US, it is common for children and teenagers to go away on adventure holidays for some time each summer. Nursery nurses are much in demand for short periods of employment (usually three months) on these schemes. The pay is low, especially in comparison with US rates of pay (around £80-£110 per week), and accommodation more basic than that available when working for a package tour company. You do, however, have a free flight and can delay your departure. Many nursery nurses use this as a way of covering costs for an extended tour of the US or Canada.

Further Education

There is a health diploma for nursery nurses as well as a diploma in post-qualifying studies which allows nursery nurses to specialise. The SNNB post-certificate award can be studied over one year part time for nursery nurses with three years' experience. There is a certificate in play-skills, a one-year course which awards a special qualification for nursery nurses working with children between the ages of two and a half and five years.

Nursery nurses can also undertake teaching qualifications which range from the City & Guilds 730 teaching certificate to a degree in teaching (BEd). These will enable nursery nurses to teach NNEB courses themselves.

Career Options

Nursery nursing is seen by some careers advisers as lacking a proper career path but it offers a wealth of options for qualified nursery nurses who can work in a range of settings and can themselves become lecturers in colleges of further education teaching NNEB courses.

Further Information

Information about training can be obtained from:

The National Nursery Examination Board, 8 Chequer Street, St Albans, Hertfordshire AL1 3XZ

The Association of Nursery Training Colleges, Secretary: Mrs M Randle, The Princess Christian College, 26 Wilbraham Road, Fallowfield, Manchester M14 6JX

Scottish Nursery Nurses Board, 6 Kilnford Crescent, Dundonald, Kilmarnock, Ayrshire KA2 9DW

These institutions produce leaflets called *Becoming a nursery nurse* and *Nursery work-job outlines*.

Useful Reading
Posts are advertised in *The Lady*, *Nursery World*, the *Guardian* and the *Times Educational Supplement*; local vacancies are advertised in local newspapers.
Careers Working with Children and Young People, Kogan Page.

Chapter 11
Careers Related to Nursing

Operating Department Assistant

This job involves assisting in a hospital operating theatre. The assistant works as a member of a team which includes the surgeon, anaesthetist and a trained nurse. One of the assistant's tasks is to provide extra support in the management of the complicated equipment now used in operating theatres.

Operating department assistants must be physically fit, quick and competent, and able to work as part of a team.

Entry

Candidates must be over 18, have four GCSEs or equivalent, or sit an entrance test. An aptitude for dealing with equipment is useful. Training is on the job and by attendance at classes and training courses. Training lasts for two years and leads to City and Guilds examinations.

Starting Salary (1994)

Trainees: about £7,000; about £9,000 when qualified.

Further Information

British Association of Operating Department Assistants, 70A Crayford High Street, Dartford, Kent DA1 4EF.

Ambulance Work

Ambulance work involves three types of ambulance staff and also clerical (control room) assistants. Ambulance care assistants transport ill, immobile or elderly patients to and from hospitals, clinics and day hospitals. Ambulance technicians are involved in a full range of duties including emergency work. Paramedics are experienced technicians who have undertaken further specialised training. Their work involves an extended range of duties including siting intravenous infusions (drips) and defibrillation (applying an electric shock to the chest wall to restart a heart or restore normal rhythm). They also intubate (put a special tube into the throat to allow a patient to breathe), administer drugs and assist at major accidents. Control room assistants answer emergency calls and send ambulance personnel to the emergency. They operate switchboards and use computers in their work.

There are 66 ambulance service divisions which are currently run as independent NHS Trusts. All have slightly different entry criteria and training schemes so the information here is only a general outline.

Entry

In some ambulance services, all entry is via the ambulance care assistant route; in others, it is possible to enter as an ambulance technician but not as a paramedic. The usual entry age is 18 years but entry is sometimes possible via a cadet scheme at 16 or 17. Entrants (with the exception of cadets) are required to have a full, clean, UK driving licence and to be physically fit as the work involves driving and lifting.

Training

Ambulance care assistants have a four-week training which is run by the local ambulance brigade. Technicians have an eight-week residential course; six weeks are spent on patient care and two weeks on advanced driving techniques. There is a further year of supervised practice with an experienced ambulance technician and assessments are carried out every three months. A National Ambulance Proficiency certificate is awarded on successful completion of training. Following a further year's experience, technicians can apply for paramedic training which lasts four weeks. After successful completion of training, students spend a further four weeks training in hospital under the supervision of a hospital consultant.

Control room assistants can, in some services, progress through management grades, but in other services they must have a period of clinical experience as a technician before they can be considered for promotion.

Career Prospects

Career prospects are excellent and there are opportunities to move into management or teaching. The majority of senior ambulance personnel began their careers working in ambulances. Careers in ambulance work are open to men and women but until recently most ambulance personnel were men. This is now changing and it is not unusual for two women to work together on emergency or paramedic vehicles.

Chiropody

Chiropodists are trained to diagnose and treat foot problems and to advise patients on how to prevent foot problems in the future. They work mainly with 'priority groups' within the NHS. Priority groups are children, pregnant women and elderly people.

The chiropodist deals with minor foot problems such as corns, callouses, bunions and blisters. She also deals with major problems which can be present at birth or develop as a result of illness. Her work consists of exercise, massage, applying dressings, prescribing drugs and creams and undertaking foot surgery such as removing ingrowing toe nails. Health education is an important part of her role and she often gives talks on foot care to school children, to pregnant women at ante-natal classes and to elderly people at pensioners' groups.

Training
All students must be at least 18 years old and have two A or H levels and at least three GCSEs at grade A-C or S grades 1-3. Some colleges ask for A levels in specific subjects such as biology or another science subject, and in Scotland H level maths is required. Mature entrants are welcomed and most colleges will waive their entry requirements if the student has relevant experience, such as having worked as a foot-care assistant. (Foot-care assistants assist chiropodists in clinical work.)

The courses last three years full time and include examinations throughout the course and a final written examination. Successful candidates are entitled to call themselves state registered chiropodists.

Pay and Conditions in the NHS
Newly qualified NHS chiropodists earn approx £13,120 per year for a 37.5 hour week. Chiropodists in senior posts can earn up to £29,000 per year (1996 figures) but there are few senior posts available. London weighting is paid in addition to the above salaries to chiropodists working in London. Details of London weighting are given on page 28.

Work in Private Practice
Many chiropodists begin by working in the NHS but leave to work in private practice. Chiropodists in private practice often work longer hours than their NHS counterparts and usually work at weekends and in the evening. They may work part time or on a sessional basis. Chiropodists in private practice can earn much more than in the NHS.

Part-time Work
There are many opportunities for chiropodists to work part-time within the NHS. The NHS frequently employs chiropodists on a 'sessional basis'. The chiropodist is paid for three-hour sessions in hospitals, clinics, GPs' surgeries or in the community. She sees a number of patients during this time. Some chiropodists in private practice also do sessional NHS work during the day and private work at evenings and weekends.

Further Information
Institute of Chiropodists, 91 Lord Street, Southport, Merseyside PR8 1SA

Society of Chiropodists, 53 Welbeck Street, London W1M 7HE

Dentistry
There are a number of careers in dentistry: the dental technician, dental surgery assistant and dental hygienist.

Dental Technicians

Dental technicians make dental appliances up from prescriptions given to them by dental surgeons. This work is extremely skilled and is becoming more technical as advances are made in dental technique. The work includes making full and partial dentures; making crowns and bridges from porcelain, acrylic or metal; making appliances such as braces from a variety of materials; making prostheses which help injured or ill patients to eat or speak normally.

Trainees are required to have excellent colour vision, essential for matching crowns and bridges to the patients' own teeth. They are also required to have a minimum of five GCSEs or S levels including a science subject. Entry requirements can be waived for mature entrants.

There are two types of course: a hospital-based apprenticeship which lasts five years or a full-time, three-year college course.

The work in either NHS or private practice is carried out in laboratories and there is little opportunity for patient contact.

Qualified dental technicians' salaries vary from region to region but range from £9,000 to £28,000 per annum depending on grade (1996 figures).

Further Information

The Dental Laboratories Association, Chapel House, Noel Street, Nottingham NG7 6AS

Dental Technician Education and Training Advisory Board, c/o British Dental Association, 64 Wimpole Street, London W1M 8AL

General Dental Council, 37 Wimpole Street, London W1M 8DQ

National Joint Council for the Craft of Dental Technician, 64 Wimpole Street, London W1M 8AL.

Dental Surgery Assistant

Dental surgery assistants work with dentists and hygienists. The work involves clinical and clerical duties. In a dentist's surgery, she may be responsible for record-keeping and sending out reminders and follow-up appointments; in larger surgeries where several dentists work, these duties may be undertaken by a receptionist. She assists the dentist while he is carrying out treatments and may develop and mount dental X-rays. She may prepare materials used in treatment. She also has an important role in reassuring nervous patients.

Many dentists are willing to train DSAs without formal qualifications. There are also courses available in hospital or dental schools. These are full-time, one- or two-year courses and a salary is paid during training. Students are usually required to have between two and four GCSEs at grades A–C and can commence training at age 16 or 17 (the starting age varies between schools). There are practical, oral and written examinations which lead to the National Certificate in Dental Surgery.

Many advanced courses are available for experienced DSAs and these are usually run at dental schools.

After a period of experience, usually a minimum of one year, DSAs can undertake further training to become dental hygienists, DSA teachers or senior DSAs with responsibility for supervising other DSAs, usually within a hospital or dental school department.

Further Information

Association of British Dental Surgery Assistants, DSA House, 29 London Street, Fleetwood, Lancashire FY7 6JY.

Dental Hygienists

Dental hygienists prevent tooth decay and educate patients to care for their teeth. The work of the dental hygienist involves scaling, polishing and cleaning teeth and removing stains. She also applies fluoride and special sealing material to teeth to reduce the risk of decay. She scales and polishes teeth when patients have been unable to clean or have their teeth cleaned because of injuries, such as a broken jaw.

In the UK, dental hygienists work under the supervision of a dental surgeon but in other countries they are practitioners in their own right. This supervision is not rigorous and dental hygienists are keen to work as practitioners in their own right. It is not possible for patients in general practice to book appointments with a hygienist without first seeing a dentist. It is common for hospital doctors to refer patients to hygienists for treatment.

The hygienist works in many areas: hospital, clinics, the community, occupational health, dental surgeries, the armed forces, teaching and research.

Many dental hygienists were previously DSAs who undertook further training. Most colleges require hygienists to have a minimum of one to two years' DSA experience and many demand the National Certificate in Dental Surgery (see under DSA above). Colleges usually require five GCSEs at grade A-C or S grades 1-3 including English language and either biology or human biology. The course is a two-year, full-time, college-based course and leads to the Certificate of Proficiency in Dental Hygiene. Students must be at least 20 years old.

On qualification, hygienists can choose to work either in hospitals or dental schools where there are career opportunities. After further training they can become tutors or manage other hygienists. There are few opportunities for career development within general practice at present but work within general practice often appeals to those who wish to work part time.

There are opportunities for hygienists to work part time on a sessional basis in hospital and general practice. Dental hygiene is a new and fast-growing profession; it was unknown before the early 1970s but is becoming more popular because of the increased emphasis on preventative dental work. The majority of hygienists are under 40 years old and work part time.

NHS salaries vary. Each NHS Trust tends to pay a different rate. The starting salary is approximately £12,000, rising to around £17,000 after five years' experience. Salaries are higher in general practice.

Further Information

The British Dental Hygienists' Association, 13 The Ridge, Yatton, Bristol BS19 4DQ.

Dietetics

The links between ill-health and poor diet are now well proved and there has been much research in recent years on how special diets can actually help individuals to regain good health. Dieticians have two main roles: to treat ill health by special diets and to promote good health through a healthy diet.

Dieticians normally work in a number of areas: caring for patients during their stay in hospital and as hospital outpatients, working in the community advising groups specially in need of the best possible diet such as pregnant women, mothers of school children and elderly people; advising people with nutritional disorders such as obesity and underweight how to deal with these problems; undertaking research on the links between diet and specific diseases such as arthritis; working for a food company advising on the content of foods.

Entry

Minimum entry requirements are two A levels or H levels. Some colleges prefer science subjects such as biology or chemistry while others specify maths. Mature entrants are considered and the course requirements can be waived but most colleges demand evidence of recent study. This is to ensure, as far as possible, that students can cope with the academic demands of the course. The course is college based but includes periods of practical experience. It is normally four years full time and leads to a degree in dietetics and state registration as a dietician. There are two-year, postgraduate diplomas available for graduates with relevant degrees; subjects classed as relevant include physiology and biochemistry.

Pay and conditions

Salaries for newly qualified dieticians commence at approx £13,120 per year. Senior dieticians earn from £17,500 to £20,000 per year and managers up to £29,000 per year (1996 figures). London weighting allowance is paid to staff working in London. There are additional allowances for training and 'on-call' duties.

Working hours are normally a 36-hour week, Monday to Friday. Dieticians working in hospitals may have to work some evenings, weekends and be 'on call'.

Career Opportunities

There are opportunities for promotion within the NHS via the normal career path; the dietician also has the opportunity to become a lecturer in dietetics or an NHS manager.

Further Information

The British Dietetic Association, 7th Floor, Elizabeth House, 22 Suffolk Street, Queensway, Birmingham B1 1LS.

Nurse's Aide

In the past student nurses provided differing levels of skilled care to patients under the supervision of trained nurses. Now that student nurses are not considered part of the work force there is a need for a support worker or health care assistant who can provide more highly skilled care than the traditional nursing auxiliary, under the supervision of trained nurses.

Many nursing auxiliaries developed their skills because they had the good fortune to be placed on a ward or department where nursing staff spent time training them. Their training has been much less organised and detailed than that of health care assistants.

Health care assistants will undertake a period of training. Some of the training will take place in either a local nursing college or a college of further education but much of it will involve learning in the work environment, in much the same way as in traditional registered nurse training programmes. This training will lead to the award of National Vocational Qualifications (NVQs).

There are currently four levels of skill recognised under the NVQ scheme and the government has given approval for NVQs to extend beyond level 4 into the professional domain. It is possible that at some time in the future registered nursing qualifications will be linked with higher level NVQs. This is similar to the way recently discontinued enrolled nurse training developed and the NVQ courses may appeal to those who would like to gain practical nursing skills and would have previously been attracted to enrolled nurse training.

Nursing auxiliaries will be able to undertake HCA training to gain NVQs. It is thought that eventually nursing auxiliary posts will be phased out and replaced by HCAs with different levels of skills.

Case Studies

A health care assistant talks about her work. *Janet* is 19 and is training as a health care assistant at her local hospital. She describes her work.

> I didn't know what I wanted to do when I left school and had several jobs which I didn't really like. I wanted to work with people. I didn't apply to do nurse training as I'm a practical person and didn't want to commit myself to years of study. The HCA role interested me because I'm learning all the time but I'm also working. I feel that it's not just a dead-end job like being a nursing auxiliary. I will have qualifications which will be recognised if I move to another area.
>
> I enjoy the work. I help the patients to wash and bath when they are ill. I have learnt how to take temperatures and blood pressures. I have learnt how to

admit patients. I help the nurses with bed making and other tasks and can always rely on the help, support and supervision of trained nurses.

Judith Capper has worked as a care assistant for over 15 years. She is currently studying for her NVQ level 3 at the local college of further education.

> I've been doing many of the things on the course for some years now but I am studying in greater depth than before. I am learning 'why' I have been taught to do things in certain ways. The NVQ course gives me the opportunity to gain a formal qualification and recognises the skills I have acquired over the years.

Judith has been promoted to senior care assistant at work and helps with much of the paperwork. This enables registered nurses to spend more time with the patients.

Occupational Therapy

The role of the occupational therapist has changed beyond all recognition in recent years. Once occupational therapy was thought of as therapy which prevented patients from becoming bored while they were in hospital for long periods of treatment. That now forms a small part of the OT's work. Much of the work involves people who need to learn or re-learn skills which will enable them to care for themselves on their return home. OTs are involved in assessing patients' abilities to carry out the activities of daily living. The work might involve helping a patient learn to dress and care for herself again following a stroke. The OT may then go home with her patient to see if any adaptations are required to enable the patient to return home and live as independent a life as possible. OTs work in general hospitals, in units caring for people with learning disabilities, in day hospitals catering for elderly people and with young disabled clients.

Entry and Training

Most colleges require two A or H levels though entry requirements can be waived for mature entrants, especially those with experience in working as an occupational therapy aide. Most colleges will not accept students over 40 years old. The minimum age for entry is 18. The course lasts four years and after successful completion students are awarded a degree in occupational therapy and become state registered occupational therapists. The course is college based but one-third of it is practical work under the supervision of a registered occupational therapist.

Pay and Conditions

Salaries start at around £13,120 per year for newly qualified OTs; more senior management posts pay salaries of around £29,000 per year (1996 figures). London weighting is paid to OTs who work in London. OTs normally work a 36-hour week over five days. There may be occasional weekend or 'on-call' work but this is unusual. There are many opportunities for part-time work and job sharing is more established than in other professions.

Career Opportunities
Most OTs work within the NHS but there are opportunities to work in other areas such as social services' residential homes or for charities providing care for disabled children. There are opportunities for OTs to become managers or teachers within the NHS, but few similar opportunities outside the NHS at present though this may well change as social services become responsible for community care.

Further Information
The College of Occupational Therapists and the British Association of Occupational Therapy share the same address:

6-8 Marshalsea Road, Southwark, London SE1 1HL.

Physiotherapy

There are currently four applicants for every available training place; this makes physiotherapy a more popular career choice than medicine.

Physiotherapists are commonly known as 'physios'. The physio works with individuals who have diminished or absent function because of congenital disease, acquired disease or because of an accident. She aims, where possible, to work with her patient to restore normal function. When it is not possible to restore function, the physio works with patients to maximise ability and minimise disability.

Physios aim to get patients to do the work of exercising and restoring function to weak areas. This is more effective than passive movements in most cases. Physios work with people of all ages; newborn babies, children, women who have just given birth, middle-aged people suffering from arthritis and with elderly people who have suffered fractures or strokes. The majority of work is on a one-to-one basis and not in groups.

Entry and Training
The usual entry qualifications are two A or H levels and the preferred subjects are physics, chemistry and biology. Entry requirements can be waived for mature entrants but as competition is so intense only individuals with relevant experience such as work as a physiotherapy aide are considered by many colleges. The number of peope entering physiotherapy as mature students is increasing each year and in 1995 more than 20 per cent of physios in training were mature students.

All candidates must have a medical examination to ensure that they are physically fit before acceptance. Some colleges have a minimum height requirement. Training takes place at university and is a first degree. After this course, successful students become state registered physiotherapists and join the Chartered Society of Physiotherapists.

Pay and Conditions
Salaries in the NHS start at £13,120 for newly qualified physiotherapists and rise to around £29,000 for senior staff (1996 figures). London weighting is paid to staff who work in London.

NHS physios work in hospital or the community. The majority of posts are currently in hospitals but this is changing as the main focus of care moves from hospital to community.

Hospital-based physios work 35 hours per week over five days. All physios work a set number of weekends and undertake some evening work as they provide emergency services such as chest physiotherapy to acutely ill patients. There are also 'on-call' duties.

There are opportunities for part-time work and some health authorities employ community physios on a sessional basis. Many posts are available in community physiotherapy in some parts of the country as many physios prefer hospital-based work.

About 15 per cent of all physios now work in private practice. A few work in research and although the numbers undertaking research are small, this is a growing area.

Career Opportunities

There are opportunities within the NHS for physios to specialise in particular areas, become managers, to teach or to undertake research. There are few opportunities for career development outside the NHS.

Further Information

The Chartered Society of Physiotherapists, 14 Bedford Row, London WC1R 4ED

The North London School of Physiotherapy for the Visually Handicapped, 10 Highgate Hill, London N19 5ND

Queen's College Glasgow, Department of Physiotherapy, 1 Park Drive, Glasgow G3 6LP.

Speech and Language Therapy

Speech and language therapists were known as speech therapists until recently. The work involves treating patients with difficulties in speech and language. These difficulties may be caused by a congenital condition such as a cleft palate or as the result of disease, such as a stroke. Most of the work is carried out on an one-to-one basis and involves working closely with other therapists, nurses and medical staff.

Entry and Training

The usual entry requirements are two A levels or equivalent and English at A level is often required. Entry requirements are often waived for mature entrants. Training is university based and is three years for a basic degree and four years for an honours degree. Graduates in a relevant subject such as psychology can apply to undertake a two-year postgraduate diploma. Both courses combine theoretical and practical work. These courses enable speech and language therapists to register as qualified practitioners.

There are opportunities after qualification to specialise in different areas of speech therapy.

Pay and Conditions

Salaries start from £13,120 for a newly qualified therapist rising to £29,000 for management grades (1996 figures). London weighting is paid to therapists who work in London. Work is normally Monday to Friday on a 9-5 basis. There are opportunities for part-time work.

Career Opportunities

The majority of therapists are employed within the NHS where they can work in hospital or in the community. There are opportunities to become a manager or a teacher.

Some therapists supplement their income by private practice but it is doubtful if anyone could earn a reasonable income solely by working privately.

Further Information

The Royal College of Speech and Language Therapists, 7 Bath Place, Rivington Street, London EC2A 3DR

Radiography

Radiographers normally work in hospital. There are two types of radiographer.

Diagnostic radiographers are responsible for taking X-rays, carrying out ultrasound and more specialised scans to diagnose suspected problems such as a broken bone or a brain tumour.

Therapeutic radiographers are responsible for administering radiation to patients as a treatment. This treatment is prescribed by doctors for diseases such as an overactive thyroid gland or cancer. In many cases, the same therapeutic radiographer will administer a series of treatments over a long period to build up a relationship with her patient.

Entry and Training

The normal entry requirements are two A or H levels and specific subjects such as physics or biology may be required. All courses are now at degree level and are held at universities.

Pay and Conditions

Diagnostic radiographers work in the X-ray department and carry out mobile X-rays on seriously ill patients. Therapeutic radiographers work in radiotherapy departments and work closely with other specialists.

All radiographers normally work a 35-hour week over five days. Diagnostic radiographers work at weekends and on bank holidays and are 'on-call'. Therapeutic radiographers tend to work Monday to Friday with bank holidays off.

Salary scales start at £13,120 and rise to £29,000 per year (1996 figures). Additional allowances are paid for unsocial hours, 'on-call' and working in London.

Career Prospects
There are opportunities for radiographers to become head of department or, after further training, to become teachers.

Social Work

Social workers work in a variety of settings: in hospital; in social services departments of local authorities; children's homes; or in residential homes for the elderly.

There are different types of social worker. The medical social worker is usually based in hospital and works with occupational therapists, physiotherapists, doctors and nurses to assess a patient's needs on discharge after illnesses such as a stroke.

The psychiatric social worker works with community psychiatric nurses and doctors in supporting people recovering from mental illness and her work can be hospital or community based. The 'generic' or non-specialised social worker is usually part of a team based in the local authority's social services department, and works with a variety of clients.

Entry and Training
Some social workers are unqualified but the majority now hold some form of recognised qualification such as the Certificate of Qualification in Social Work (CQSW) or the Certificate in Social Services (CSS). There have been many changes in the training of social workers in recent years and the basic qualification is now the two-year Diploma in Social Work.

The Diploma in Social Work (DipSW) has replaced all previous qualifying awards. This is a two-year full-time course. It may be studied on a part-time day-release basis or by distance learning. The Diploma combines practical and academic training. It can be combined with an academic award such as the Diploma in Higher Education (DipHE), Bachelors Degree, Postgraduate Diploma or a Masters Degree. Programmes are planned and run jointly by universities and social work agencies. All education programmes now begin at the start of the academic year in October.

The DipSW is a generic qualification. Students have to acquire a general understanding and meet academic and practice standards for six core competencies. Students can follow either a general or specialised pathway. Specialist pathways include training to work in a particular service such as criminal justice, or training to work with a particular client group such as children.

Pay and Conditions
Salaries commence at £13,120 for basic grade social workers and can rise to as much as £80,000 for directors of social services in a local authority (1996 figures). Many social workers work 9–5 Monday to Friday but there is an element of weekend working and there is always a social worker on duty outside office hours, during the night, at weekends and over holiday

periods. This is usually on a rota basis. There are opportunities for part-time work, and job sharing is well established in local authorities.

Career Opportunities

There are many opportunities within management and in teaching for social workers. Senior management posts are well paid.

Further Information

Central Council for Education and Training in Social Work (CCETSW) produce a package giving details of all social worker training courses and entry criteria.

England:
Information Service, Derbyshire House, St Chad's Street, London WC1H 5AD

Scotland:
78-80 George Street, Edinburgh EH2 3BU

Wales:
Head Office, West Wing, St David's House, Wood Street, Cardiff CF1 1ES.

See *Careers in Social Work*, published by Kogan Page.

Therapy Aides

In recent years there have been shortages of qualified chiropodists, occupational therapists and physiotherapists and so these therapists have taken on assistants to help them with the more routine aspects of their work.

Training is on a one-to-one basis; there are no formal qualifications but most therapy aides have a wealth of experience. There are no prospects of promotion but some aides go on to become mature students (who do not usually require formal entry qualifications) and eventually qualify as therapists.

Therapy aides posts are usually advertised in local newspapers. It may be worth contacting your local hospital to see if they have any posts available.

Chapter 12
Training Courses

Key: A = Adult Nursing
AGP = Accelerated Graduate Programme
C = Children's Nursing
L = Learning Disability Nursing
M = Mental Health Nursing

Pre-Registration Courses Available in England at Approved Institutions

Anglia Polytechnic University, Faculty of Health & Social Work, Admissions/Placement Department, Churchfield One, Southend Hospital Prittlewell Chase, Westcliffe on Sea, Essex SS0 0RY. Tel: 01702 335562.
Diploma: A, C, L, M. No AGP. Degree: A, C, L, M.

Bloomsbury & Islington College of Nursing & Midwifery, Minerva House, 1-4 North Crescent, Chenies Street, London WC1E 7ER. Tel: 0171 387 9300.
Diploma: A, M. No AGP, no degree.

Bournemouth University, Institute of Health Services, Bournemouth House, Christchurch Road, Bournemouth BH1 3LG. Tel: 01202 524111.
Diploma: A, C, L, M. No AGP. Degree: A, M (3 years).

Buckinghamshire College - A College of Brunel University, Admissions Department, Queen Alexandra Road, High Wycombe, Bucks HP11 2JZ. Tel: 01494 522141.
Diploma: A, C, M. No AGP, no degree.

Buckinghamshire College of Nursing & Midwifery, Vale of Aylesbury Centre, Stoke Mandeville Hospital, Aylesbury, Bucks HP21 8AL. *Also* Lovelock Jones Nursing Education Centre, Wycombe General Hospital, Bucks HP11 1QN. Tel: 01296 315578.
Diploma: A, C, L, M. No AGP, no degree.

Canterbury Christ Church College, The Admissions Office, Canterbury, Kent CT1 1QU. Tel: 01227 767700
Diploma: A, C, L, M. AGP: A, C, M. No degree.

Chester College of Higher Education, School of Nursing, Cheyney Road, Chester CH1 4BJ. Tel: 01244 364668.
Diploma: A, C, M. No AGP, no degree.

City University, St Bartholemew's School of Nursing & Midwifery, 20 Bartholomew Close, London EC1A 7QN. Tel: 0171 505 5720.
Diploma: A, C, M. No AGP, no degree.

Coventry University, Main Site, Walgrave Hospital, Clifford Bridge Road, Coventry CV2 2DX. Tel: 01203 538728.
Diploma: A, C, L, M. No AGP, no degree.

Training Courses 87

De Montfort University, School of Health & Community Studies, Scraptoft Campus, Leicester LE7 9SU. Tel: 0116 255 1551.
Diploma: A, C, L, M. No AGP. Degree: A, C, M.

Edgehill College of Further Education, School of Health Studies, Aintree Complex, Fazakerly Hospital, Longmoor Lane, Liverpool L9 7AL. Tel: 0151 529 3336.
Diploma: A, C, L, M. No AGP, no degree.

Homerton College, Cambridge, Education Centre, Addenbrooks Hospital, Cambridge CB2 2QQ. Tel: 01223 216248.

Recruitment Officer, Education Centre, Peterborough District Hospital, Thorpe Road, Peterborough PE3 6DA. Tel: 01733 67451 ext: 4769.
Diploma: A, C, L, M. No AGP, no degree.

Humberside College of Health, Student Services, College House, East Riding Centre, Beverly Road, Willerby, Hull HU10 6NS. Tel: 01482 675644.
Diploma: A, C, L, M. No AGP, no degree.

Keele University, Department of Nursing and Midwifery, City General Hospital, Newcastle Road, Stoke on Trent ST4 6QG.
Diploma: A, C, M. No AGP, no degree.

King Alfred's College of Further Education, School of Health and Community Studies, Admissions Officer, Sparkford Road, Winchester SO22 4NR. Tel: 01962 827346.
Diploma: A, C, M. No AGP, no degree.

King's College, The Nightingale Institute, The Registry, Cornwall House, Waterloo Road, London SE1 8WA. Tel: 0171 873 5111.
Diploma: A, C, L, M. AGP: A, C, L, M. Degree: A, M (4 years).

Kingston University & St George's Medical School, Joint Faculty of Health Sciences, Kingston University, Penrhyn Road, Kingston on Thames KT1 2EE. Tel: 01372 734 778.
Diploma: A, C, L, M. AGP: A, C, L, M. No degree.

Leeds Metropolitan University, Calverly Street, Leeds LS1 3HE.
Diploma: A, C, L, M. AGP: A, C, L, M. Degree: A, M (4 years).

John Moores University, School of Health Care, Central Applications Department, 79 Tithebarn Street, Liverpool L2 2ER. Tel: 0151 231 2121.
Diploma: A, C, M. No AGP. Degree: A, C, M (4 years).

Manchester College of Nursing & Midwifery, Recruitment Centre, Gateway House, Piccadilly South, Manchester M60 7LP. Tel: 0151 231 4127.
Diploma: A, C, M. No AGP, no degree.

University of Manchester, Department of Nursing, Manchester M13 9PT. Tel: 0161 275 5333.
Degree: A, L, M, C + Health Visiting or District Nursing (4 years).

Middlesex University, Faculty of Health Studies, Old Nurses' Home, North Middlesex Hospital, Sterling Way, London N18 1QX. Tel: 0181 887 2733.
Diploma: A, C, M. No AGP. Degree: A, C, M (3 years).

Nene Centre for Healthcare Education, Admissions, Nene College of Higher Education, Unit 3, Park Campus, Boughton Green Road, Northampton NN2 7AL. Tel: 01604 735500 ext. 2633.
Diploma: A, C, L, M. No AGP, no degree.

Oxford Brookes University, School of Health Care Studies, John Radcliffe Hospital, Headley Way, Headington, Oxford OX3 9DU. Tel: 01865 221551.
Degree: A, C, L, M (4 years).

88 Careers in Nursing and Related Professions

The Robert Gordon School of Nursing, Kepplestone Annexe, Queens Road, Aberdeen AB15 4PH. Tel: 01224 263360.
Diploma: A, C, L, M. No AGP, no degree.

Sheffield Hallam University, Department of Health Studies, 36 Collegiate Crescent, Sheffield S10 2BP. Tel: 0114 272 0911.
Degree: A, C, M (3 years).

South Bank University, Redwood College of Health Studies, 103 Borough Road, London SE1 0AA. Tel: 0171 928 8989.
Degree: A, L, M (3 years).

South Bank University, School of Paediatric Nursing & Child Health, Faculty of Health & Social Care, 103 Borough Road, London SE1 0AA. Tel: 0171 928 8989.
Diploma: C. No AGP. Degree: L and Social Work (4 years).

Southampton University, School of Nursing & Midwifery, South Academic Block, General Hospital, Tremona Road, Southampton, Hants SO16 6YD. Tel: 01703 796550.
Diploma: A, C, L, M. No AGP. Degree: A, C (4 years).

Staffordshire University, School of Health Education, Blackheath Lane, Stafford ST18 0AD. Tel: 01785 353794.
Diploma: A, C, L, M. No AGP, no degree.

University of Hull, Institute of Nursing Studies, Hull HU6 7RX.
Degree: A, L (4 years).

University of Leeds, School of Health Care Studies, St James University Hospital Site, Beckett Street, Leeds LS9 7TF.
Diploma: A, C, L, M. AGP: A. No degree.

University of Liverpool, Department of Nursing, The Whelan Building, PO Box 147, Liverpool L69 3BX.
Degree: A, A + Health Visiting or District Nursing (4 years).

University of Luton, Faculty of Health Care & Social Studies, Admissions Office, Park Square, Luton LU1 3JU. Tel: 01582 489218.
Diploma: A, C, L, M. No AGP, no degree.

University of Northumbria, Faculty of Health, Social Work & Education, Teaching Centre, Freeman Hospital, High Heaton, Newcastle upon Tyne NE7 7DN. Tel: 0191 264 3111 ext 26078.
Diploma: A, C, L, M. No AGP. Degree: M (3 years).

University of Nottingham, School of Nursing & Midwifery, Centralised Admissions, B Floor, Queens Medical Centre, Nottingham NG7 2UH. Tel: 0115 924 9924.
Diploma: A, C, L, M. AGP: A, C, L, M. Degree: A, C, L, M (4 years).

University of Plymouth, Institute of Health Studies, Drake Circus, Plymouth, Devon PL4 8AA. Tel: 0345 626553.
Diploma: A, C, L, M. No AGP, no degree.

University of Portsmouth, School of Health Studies, Queen Alexandra Hospital, Cosham, Portsmouth PO6 3LY. Tel: 01705 379451.
Diploma: A, C, L, M. No AGP, no degree.

University of Salford, Department of Nursing & Midwifery, Student Recruitment Office, 2nd Floor, Peel House, Albert Street, Eccles, Manchester M30 0NJ. Tel: 0161 952 2313.
Diploma: A, C, M. No AGP, no degree.

University of Sheffield, School of Nursing & Midwifery, Admin Headquarters, Faculty of Medicine, Beech Hill Road, Sheffield S10 2RX. Tel: 0114 271 1910.
Diploma: A, C, L, M. No AGP, no degree.

University of Surrey, European Institute of Health & Medical Sciences, PO Box 251, Guildford, Surrey GU1 3ZX. Tel: 01483 464162.
Diploma: A, C, L, M. No AGP. Degree: A, C, M (4 years).

University of Teesside, College of Health, Recruitment Department, Education Centre, South Clevelands Hospital, Marton Road, Middlesbrough, Cleveland TS4 3BW. Tel: 01642 854814 ext 3697.
Diploma: A, C, L, M. No AGP, no degree.

University of the West of England, Admissions Office, Frenchay Campus, Coldharbour Lane, Bristol BS16 1QY. Tel: 0117 965 6261.
Diploma: A, C, L, M. No AGP, no degree.

University of Wolverhampton, School of Nursing & Midwifery, The Recruitment Office, Manor Hospital, Moat Road, Walsall, West Midlands WS2 9PS.
Diploma: A, C, L, M. No AGP, no degree.

University of York, Dept of Health Studies, Innovation Centre, Science Park, York YO1 5DG. Tel: 01904 435222.
Diploma: A, C, L, M. No AGP, no degree.

West Yorkshire College of Health, The Registry, Lee House, Stanley Royd Hospital, Aberford Road, Wakefield WF1 4DH. Tel: 01942 814879.
Diploma: A, C, L, M. No AGP, no degree.

Worcester College of Higher Education, School of Health Studies, Nursing & Midwifery, Henwick Grove, Worcester WR2 6AJ. Tel: 01905 855000.
Diploma: A, M. No AGP, no degree.

Diploma-level Pre-Registration Nursing Programmes in Northern Ireland

Eastern Area College of Nursing Northside, Director of Nurse Education, Musson House, Grosvenor Road, Belfast BT12 6BA. Tel: 01232 263238.
Diploma: A, C. *Note.* Age limits 17 years 6 months to 45 years. A levels preferred.

Eastern Area College of Nursing Southside, Director of Nurse Education, Knockbracken Healthcare Park, 301 Saintfield Road, Belfast BT8 8BH. Tel: 01232 401922.
Diploma: A, M.

Northern Area College of Nursing, Director of Nurse Education, Antrim Hospital Site, Bush Road, Antrim BT41 2QB. Tel: 01849 424200.
Diploma: A, L, M.

Southern Area College of Nursing, Director of Nurse Education, 68 Lurgan Road, Craigavon, Co Armagh BT63 5QQ. Tel: 01762 334444.
Diploma: A, C, L, M. *Note.* Limited number of places on child branch in association with Eastern Area College Northside.

Western Area College of Nursing, Director of Nurse Education, Multidisciplinary Education Centre, Altnagelvin Hospital, Londonderry BT47 1SB. Tel: 01504 45171.
Diploma: A, C, L, M. *Note.* Will offer DC test to individuals over 21.

Diploma-level Programmes in Wales

Admissions Office, Department of Nursing, Midwifery and Health Care, University of Wales, Parc Beck Campus, Sketty Road, Swansea SA2 9DX. Tel: 01792 280647.
Programmes: A, C, M.

Mrs G Jayne, Applications Officer, School of Nursing & Midwifery, University of Glamorgan, Treforest, Pontypridd CF3 1DL. Tel: 01443 482639.
Programmes: A, C, L, M.

Director of Pre-Registration Studies, School of Nursing & Midwifery Studies, Faculty of Health, University of Wales, Fron Heulog, Ffriddoed Road, Bangor, Gwynedd LL57 2EF. Tel: 014248 354036.
Programmes: A, C, L, M.

Miss D Hobbs, Recruitment Department, School of Nursing Studies, University College of Medicine, The Grange, Velindre Road, Whitchurch, Cardiff CF4 7XP. Tel: 01222 529287.
Programmes: A, C, M.

Diploma-level Programmes in Scotland

Bell College of Technology, Hartwood Hospital, Shotts, Lanarkshire ML7 4LA. Contact Elizabeth Miller. Tel: 01501 821514.
Nursing: A, M. Midwifery: Yes; at Lanarkshire site only. *Note.* Bell College has two sites which are to be integrated on the Hamilton campus in September 1998.

Bell College, Crichton Hall, Glencaple Road, Dumfries DG1 4TG. Contact Caroline McKinnon. Tel: 01387 244234.
Nursing: A, M. No midwifery.

University of Dundee, Forth Avenue, Kirkcaldy, Fife KY2 5YS. Contact: Miss Grace M Clarke. Tel: 01592 268888.
Nursing: A, L, M. Midwifery. The university is operating on two sites during 1996-97.

University of Dundee, Tayside College, Ninewells, Dundee DO1 9SY. Contact: Miss Eleanor Forbes, Principal. Tel: 01382 632304.
Nursing: A, C, L, M. Midwifery.

Glasgow Caledonian University, 110 St James Road, Glasgow G4 0PS. Contact: Admissions Officer. Tel: 0141 552 1562.
Nursing: A, C, L, M. Midwifery.

Napier University, 219 Colinton Road, Edinburgh. Contact: Mr George Petrie. Tel: 0131 536 5645.
Nursing: A, C, L, M. Midwifery. *Note.* Two-year shortened courses available for graduates in all branches. Combined conversion course and diploma available.

University of Paisley, Argyll & Clyde College, Corsebar Road, Paisley, Renfrewshire PA2 9PN. Contact: Mr J L Rae. Tel: 0141 887 9111.
Nursing: A, L, M. Midwifery. *Note.* Argyll & Clyde and Ayrshire & Arran Colleges amalgamated in September 1996 to form the Department of Nursing & Midwifery at the University of Paisley. Each institute will run their own courses until September 1997 when they will integrate.

University of Paisley, Ayrshire & Arran College of Nursing & Midwifery, Crosshouse Hospital, Crosshouse, Kilmarnock KA2 0BE. Contact: Mrs E Kennedy. Tel: 01563 577508.
Nursing: A, L, M. Midwifery. See note above.

The Robert Gordon University, Westburn Road, Aberdeen AB9 2X5. Contact: Mr T J Moore. Tel: 01224 681818 ext 52548.
Nursing: A, C, L, M. Midwifery. *Note.* Now incorporates the former Fosterhill College which is part of the university campus.

University of Stirling, Forth Valley College, Westburn Avenue, Falkirk FK1 5ST. Contact Mr J Gavin. Tel: 01324 635091.
Nursing: A, L, M. Midwifery. *Note.* The University of Stirling assumed responsibility for Nursing and Midwifery courses at Forth Valley and Highland & Western Isles Colleges in September 1996. Courses are currently being redesigned. Applicants should contact the colleges to check availability in autumn 1997.

Highland & Western Isles College of Nursing & Midwifery, Inverness Campus, Raigmore Hospital, Inverness IV2 3UJ. Contact Mrs Lorna Gordon. Tel: 01463 704000 ext 4214.
Nursing: A, L, M. Midwifery.

Midwifery Programmes in England at Approved Institutions

Key: DM = Diploma in Midwifery
Degree CM = Degree combined with Midwifery

Anglia Polytechnic University, Faculty of Health & Social Work, Admission/Placement Department, Churchfield One, Southend Hospital, Prittlewell Chase, Westcliffe on Sea, Essex SS0 0RY. Tel: 01702 335562.
Degree CM (3 years).

Bournemouth University, Institute of Health Services, Bournemouth House, Christchurch Road, Bournemouth BH1 3LG. Tel: 01202 524111.
DM. Degree CM (3 years).

Chester College of Higher Education, School of Nursing, Cheyney Road, Chester CH1 4BJ. Tel: 01244 364668.
DM.

Coventry University, Main Site, Walgrave Hospital, Clifford Bridge Road, Coventry CV2 2DX. Tel: 012035 38728.
DM.

De Montfort University School of Health & Community Studies, Scraptoft Campus, Leicester LE7 9SU. Tel: 0116 255 1551.
DM. Degree CM (4 years).

Edgehill College of Further Education, School of Health Studies, Aintree Complex, Fazakerly Hospital, Longmoor Lane, Liverpool L9 7AL. Tel: 0151 529 3336.
DM.

Keele University, Department of Nursing and Midwifery, City General Hospital, Newcastle Road, Stoke on Trent ST4 6QG.
DM.

King Alfred's College of Further Education, School of Health and Community Studies, Admissions Officer, Sparkford Road, Winchester SO22 4NR. Tel: 01962 827346.
DM.

King's College, The Nightingale Institute, The Registry, Cornwall House, Waterloo Road, London SE1 8WA. Tel: 0171 873 5111.
DM.

Leeds Metropolitan University, Calverly Street, Leeds LS1 3HE.
Degree CM (4 years).

John Moores University, School of Health Care, Central Applications Department, 79 Tithebarn Street, Liverpool L2 2ER.
Degree CM (4 years).

Manchester College of Nursing & Midwifery, Recruitment Centre, Gateway House, Piccadilly South, Manchester M60 7LP.
Tel: 0151 231 4127.
DM.

Middlesex University, Faculty of Health Studies, Old Nurses' Home, North Middlesex Hospital, Sterling Way, London N18 1QX. Tel: 0181 887 2733.
DM.

Nene Centre for Healthcare Education, Admissions, Nene College of Higher Education,

Unit 3, Park Campus, Boughton Green Road, Northampton NN2 7AL. Tel: 01604 735500 ext. 2633.
DM.

Oxford Brookes University, School of Health Care Studies, John Radcliffe Hospital, Headley Way, Headington, Oxford OX3 9DU. Tel: 01865 221551.
Degree CM (4 years).

The Robert Gordon School of Nursing, Kepplestone Annexe, Queen's Road, Aberdeen AB15 4PH. Tel: 01224 263360.
DM.

Southampton University, School of Nursing & Midwifery, South Academic Block, General Hospital, Tremona Road, Southampton, Hants SO16 6YD. Tel: 01703 796550.
Degree CM (4 years).

Staffordshire University School of Health Education, Blackheath Lane, Stafford ST18 0AD. Tel: 01785 353794.
DM.

University of Luton, Faculty of Health Care & Social Studies, Admissions Office, Park Square, Luton LU1 3JU. Tel: 01582 489218.
DM.

University of Northumbria, Faculty of Health, Social Work & Education, Teaching Centre, Freeman Hospital, High Heaton, Newcastle upon Tyne NE7 7DN. Tel: 0191 264 3111 ext 26078.
DM. Degree CM (3 years).

University of Nottingham, School of Nursing & Midwifery, Centralised Admissions, B Floor, Queen's Medical Centre, Nottingham NG7 2UH. Tel: 0115 924 9924.
DM.

University of Portsmouth, School of Health Studies, Queen Alexandra Hospital, Cosham, Portsmouth PO6 3LY. Tel: 01705 379451.
DM.

University of Salford, Department of Nursing & Midwifery, Student Recruitment Office, 2nd Floor, Peel House, Albert Street, Eccles, Manchester M30 0NJ. Tel: 0161 9522313.
DM.

University of Sheffield School of Nursing & Midwifery, Admin Headquarters, Faculty of Medicine, Beech Hill Road, Sheffield S10 2RX. Tel: 0114 271 1910.
DM.

University of Surrey, European Institute of Health & Medical Sciences, PO Box 251, Guildford, Surrey GU1 3ZX. Tel: 01483 464162.
DM.

University of the West of England, Admissions Office, Frenchay Campus, Coldharbour Lane, Bristol BS16 1QY. Tel: 0117 9656261.
DM.

University of Wolverhampton School of Nursing & Midwifery, The Recruitment Office, Manor Hospital, Moat Road, Walsall, West Midlands WS2 9PS.
Degree CM (3 years).

West Yorkshire College of Health, The Registry, Lee House, Stanley Royd Hospital, Aberford Road, Wakefield WF1 4DH. Tel: 01942 814879.
Degree CM (4 years).

Worcester College of Higher Education, School of Health Studies, Nursing & Midwifery, Henwick Grove, Worcester WR2 6AJ. Tel: 01905 855000.
DM.